© Troi Santos, NYULMC

MARTIN J. BLASER, MD, has studied the role of bacteria in human disease for more than thirty years. At NYU, he is the director of the Human Microbiome Program and is the Muriel G. and George W. Singer Professor of Translational Medicine. He served as president of the Infectious Diseases Society of America, as chair of the Board of Scientific Counselors of the National Cancer Institute, and as chair for Clinical Research at the National Institutes of Health. He co-founded the *Bellevue Literary Review* and his work has been written about in periodicals including *The New Yorker, Nature, The New York Times, The Economist, The Washington Post,* and *The Wall Street Journal.* His more than 100 media appearances include *The Today Show, Good Morning America,* NPR, the BBC, *The O'Reilly Factor,* and CNN. In 2014, Blaser was the Kinyoun Lecturer at the National Institute of Allergy and Infectious Diseases, and received the Alexander Fleming Award for Lifetime Achievement from the Infectious Diseases Society of America. He lives in New York City.

Additional Praise for *Missing Microbes*

"Readable and challenging, *Missing Microbes* provides a stimulus with which to probe existing dogma."   —*Science*

"Unlike some books on medicine and microbes, Dr. Blaser's doesn't stir up fears of exotic diseases or pandemic 'superbugs' resistant to all known drugs. He focuses on a simpler but more profound concern: the damage that modern life inflicts on the vast number of microbes that all of us, even healthy people, carry inside us at all times."
                                                —*The Wall Street Journal*

"*Missing Microbes* blazes a new trail."   —*The Huffington Post*

"An engrossing examination of the relatively unheralded yet dominant form of life on Earth."   —*Publishers Weekly* (starred review)

"Blaser's *Missing Microbes* is a masterful work of preventative health and superb science writing."   —*Booklist* (starred review)

"Dr. Blaser's credibility as a world-class scientist and physician makes this exploration of our body's microbial world particularly provocative. *Missing Microbes* will make you rethink some fundamental ideas about infection. Blaser's gift is to write clearly and to take the reader on a fascinating journey through the paradoxes and insights about the teeming world within us."
                          —Abraham Verghese, MD, author of *Cutting for Stone*

"*Missing Microbes* adds a new frontier toward understanding vastly underappreciated key contributions of the human microbiome to health and human disease. As a world leader in defining the microbiome, Dr. Blaser explains how disturbing its natural balance is affecting common conditions such as obesity and diabetes, long thought of as primarily nutrition and lifestyle related problems. Blaser's carefully and convincingly written book outlines new dimensions that need to be considered in fighting a number of common diseases and in promoting health and well-being."

—Richard Deckelbaum, Director, Institute of
Human Nutrition, Columbia University

"In a world that turns to antibiotics for every infection of the ear, sinuses, or skin, Dr. Blaser makes even the most nervous parent think twice about giving her child these ubiquitous drugs. . . . Dr. Blaser delivers a thoughtful, well-written, and compelling case for why doctors need to be more cautious about prescribing these medications and why consumers should consider alternatives before taking them."

—Nirav R. Shah, MD, MPH

"I have often wondered why kids today seem to have such a high incidence of asthma, ear infections, allergies, reflux esophagitis, and so many other conditions that I rarely saw growing up. This mystery has been solved by the pioneering work of Dr. Marty Blaser and is communicated brilliantly in *Missing Microbes*. I cannot emphasize enough the importance of this book to your own health, the health of your children and grandchildren, and to the health of our country. *Missing Microbes* is truly a must-read."

—Arthur Agatston, author of *The South Beach Diet*

"We live today in a world of modern plagues, defined by the alarming rise of asthma, diabetes, obesity, food allergies, and metabolic disorders. This is no accident, argues Dr. Blaser, the renowned medical researcher: the common link being the destruction of vital bacteria through the overuse of broad-spectrum antibiotics. *Missing Microbes* is science writing at its very best—crisply argued and beautifully written, with stunning insights about the human microbiome and workable solutions to an urgent global crisis."

—David M. Oshinsky, author of
the Pulitzer Prize–winning *Polio: An American Story*

MISSING MICROBES

HOW THE OVERUSE
OF ANTIBIOTICS
IS FUELING
OUR MODERN PLAGUES

MARTIN J. BLASER, MD

PICADOR

HENRY HOLT AND COMPANY
NEW YORK

www.picadorusa.com
www.twitter.com/picadorusa • www.facebook.com/picadorusa
picadorbookroom.tumblr.com

Picador® is a U.S. registered trademark and is used by Henry Holt and Company under license from Pan Books Limited.

For book club information, please visit www.facebook.com/picadorbookclub or e-mail marketing@picadorusa.com.

Designed by Meryl Sussman Levavi

The Library of Congress has cataloged the Henry Holt edition as follows:

Blaser, Martin J.
    Missing microbes : how the overuse of antibiotics is fueling our modern plagues / Dr. Martin Blaser.
        p. cm.
    Includes index.
    ISBN 978-0-8050-9810-5 (hardcover)
    ISBN 978-0-8050-9811-2 (e-book)
    1. Antibiotics. 2. Antibiotics—Effectiveness. 3. Drug resistance in microorganisms. I. Title.
    RM267.B57 2014
    615.7'922—dc23

                                                                    2013042578

Picador ISBN 978-1-250-06927-6

Picador books may be purchased for educational, business, or promotional use. For information on bulk purchases, please contact the Macmillan Corporate and Premium Sales Department at 1-800-221-7945, extension 5442, or write to specialmarkets@macmillan.com.

First published in the United States by Henry Holt and Company, LLC

First Picador Edition: February 2015

10   9   8   7   6   5   4   3   2   1

*To my children, and to future
children with a bright future*

"We live in the Age of Bacteria (as it was in the beginning, is now, and ever shall be, until the world ends) . . ."

—STEPHEN JAY GOULD, Cambridge, MA, 1993

CONTENTS

1.

MODERN PLAGUES

I never knew two of my father's sisters. In the little town where they were born, early in the last century, they didn't see their second birthdays. They had high fevers, and I am not sure what else. The situation was so dire that my grandfather went to the prayer house and changed his daughters' names to fool the angel of death. He did this for each girl. It did no good.

In 1850, one in four American babies died before his or her first birthday. Lethal epidemics swept through crowded cities, as people were packed into dark, dirty rooms with fetid air and no running water. Familiar scourges included cholera, pneumonia, scarlet fever, diphtheria, whooping cough, tuberculosis, and smallpox.

Today, only six in every thousand infants in the United States are expected to die before age one—a remarkable improvement. Over the past century and a half, our nation and other countries in the developed world have been getting healthier. Chalk it up to improved sanitation, rat control, clean drinking water, pasteurized milk, childhood vaccinations, modern medical procedures including anesthesia, and, of course, nearly seventy years of antibiotics.

In today's world, children grow up without the deformed bones caused by lack of vitamin D or "cloudy" sinuses from infections. Nearly all women survive childbirth. Eighty-year-olds, once consigned to the veranda, are swatting tennis balls, often with the help of a metal hip joint.

Yet recently, just within the past few decades, amid all of these medical advances, something has gone terribly wrong. In many different ways we appear to be getting sicker. You can see the headlines every day. We are suffering from a mysterious array of what I call "modern plagues": obesity, childhood diabetes, asthma, hay fever, food allergies, esophageal reflux and cancer, celiac disease, Crohn's disease, ulcerative colitis, autism, eczema. In all likelihood you or someone in your family or someone you know is afflicted. Unlike most lethal plagues of the past that struck relatively fast and hard, these are chronic conditions that diminish and degrade their victims' quality of life for decades.

The most visible of these plagues is obesity, defined in terms of the body mass index (BMI), which expresses the relationship between a person's height and weight. People of healthy weight have a BMI between 20 and 25. Those whose BMI is between 25 and 30 are overweight. Everyone with a BMI over 30 is obese. Barack Obama has a BMI of about 23. The BMIs of most U.S. presidents have been under 27, except for that of William Howard Taft, who once got stuck in the White House bathtub. He had a BMI of 42.

In 1990, about 12 percent of Americans were obese. By 2010, the national average was above 30 percent. Next time you go to an airport terminal, supermarket, or mall, look around and see for yourself. The obesity epidemic is not just a U.S. problem; it's global. As of 2008, according to the World Health Organization (WHO), 1.5 billion adults were overweight; of these, over 200 million men and nearly 300 million women qualified as obese. Many of these people live in developing countries that we associate more with famine than with overeating.

These figures are alarming, but the really shocking fact is that this accumulation of global human body fat has been accelerating not over the course of a few centuries but in a mere two decades. Yet fat- and sugar-rich foods, so often blamed for all the extra pounds, have been ubiquitous for a good deal longer than that, at least in the developed world, and the new generations of overweight people in the third world have not suddenly adopted a Kentucky-fried American-style diet. Epidemiologic studies have shown that high caloric intake, while definitely not helpful, is not sufficient to explain the distribution or course of the worldwide obesity epidemic.

At the same time, the autoimmune form of diabetes that begins in childhood and requires insulin injections (juvenile or Type I diabetes) has been doubling in incidence about every twenty years across the industrialized world. In Finland, where record keeping is meticulous, the incidence has risen 550 percent since 1950. This increase is not because we are detecting Type I diabetes more readily. Before insulin was discovered in the 1920s, the disease was always fatal. Nowadays, with adequate treatment, most children survive. But the disease itself has not changed; something in us has changed. Type I diabetes is also striking younger children. The average age of diagnosis used to be about nine. Now it is around six, and some children are becoming diabetic when they are three.

The recent rise in asthma, a chronic inflammation of the airways, is similarly alarming. One in twelve people (about 25 million or 8 percent of the U.S. population) had asthma in 2009, compared with one in fourteen a decade earlier. Ten percent of American children suffer wheezing, breathlessness, chest tightness, and coughing; black children have it worst: one in six has the disease. Their rate increased by 50 percent from 2001 through 2009. But the rise in asthma has not spared any ethnicity; the rates were initially different in various groups, and all have been rising.

Asthma is often triggered by something in the environment such as tobacco smoke, mold, air pollution, cockroach leavings, colds, and flu. Once an attack begins, asthmatics gasp for air and, without quick medication, are rushed to emergency rooms. Even with the best care, they can die, as did the son of a physician colleague. No economic or social class has been spared.

Food allergies are everywhere. A generation ago, peanut allergies were extremely rare. Now, if you stroll through any preschool, you will see walls plastered with "nut-free zone" bulletins. More and more children suffer immune responses to proteins in foods, not just in nuts but in milk, eggs, soy, fish, fruits—you name it, someone is allergic to it. Celiac disease, an allergy to gluten, the main protein in wheat flour, is rampant. Ten percent of children suffer from hay fever. Eczema, a chronic skin inflammation, affects more than 15 percent of children and 2 percent of adults in the United States. In industrialized nations, the number of kids with eczema has tripled in the past thirty years.

These disorders suggest that our children are experiencing levels of immune dysfunction never seen before, as well as conditions such as autism, a much discussed and debated modern plague that is a focus of my laboratory. Nor are adults escaping their own share of modern plagues. The incidence of inflammatory bowel disease, including Crohn's and ulcerative colitis, is rising, wherever we look.

When I was a medical student, esophageal reflux, which causes heartburn, was uncommon. But the ailment has exploded in these past forty years, and the cancer it leads to, adenocarcinoma of the esophagus, is the most rapidly increasing cancer in the United States and everywhere else it has been tracked, and is a particularly nasty problem for Caucasian men.

■ ■ ■

Why are all of these maladies rapidly rising at the same time across the developed world and spilling over into the developing world as it becomes more Westernized? Can it be a mere coincidence? If there are ten of these modern plagues, are there ten separate causes? That seems unlikely.

Or could there be one underlying cause fueling all these parallel increases? A single cause is easier to grasp; it is simpler, more parsimonious. But what cause could be grand enough to encompass asthma, obesity, esophageal reflux, juvenile diabetes, and allergies to specific foods, among all of the others? Eating too many calories could explain obesity but not asthma; many of the children who suffer from asthma are slim. Air pollution could explain asthma but not food allergies.

Many theories have been proposed to explain each disorder: lack of sleep makes you fat; vaccines lead to autism; genetically engineered wheat strains are toxic to the human gut; and so on.

The most popular explanation for the rise in childhood illness is the so-called hygiene hypothesis. The idea is that modern plagues are happening because we have made our world too clean. The result is that our children's immune systems have become quiescent and are therefore prone to false alarms and friendly fire. A lot of parents these days try to ramp up their kids' immune systems by exposing them to pets, farm animals, and barnyards or better still by allowing them to eat dirt.

I beg to differ. To me, such exposures are largely irrelevant to our health. The microbes present in dirt have evolved for soil, not for us. The microbes in our pets and farm animals also are not deeply rooted in our human evolution. The hygiene hypothesis, as I will show you, has been misinterpreted.

Rather we need to look closely at the microorganisms that make a living in and on our bodies, massive assemblages of competing and cooperating microbes known collectively as the microbiome. In ecology, *biome* refers to the sets of plants and animals in a community such as a jungle, forest, or coral reef. An enormous diversity of species, large and small, interact to form complex webs of mutual support. When a keystone species disappears or goes extinct the ecology suffers. It can even collapse.

Each of us hosts a similarly diverse ecology of microbes that has coevolved with our species over millennia. They thrive in the mouth, gut, nasal passages, ear canal, and on the skin. In women, they coat the vagina. The microbes that constitute your microbiome are generally acquired early in life; surprisingly, by the age of three, the populations within children resemble those of adults. Together, they play a critical role in your immunity as well as your ability to combat disease. In short, it is your microbiome that keeps you healthy. And parts of it are disappearing.

The reasons for this disaster are all around you, including overuse of antibiotics in humans and animals, Cesarian sections, and the widespread use of sanitizers and antiseptics, to name just a few. While antibiotic resistance is a huge problem—old killers like tuberculosis are increasingly resistant and making a comeback—there now seem to be separate ones, affecting people with such scourges as *Clostridium difficile* (*C. diff*), bacteria of the digestive tract resistant to multiple antibiotics, a potential danger in the hospital, and a spreading pathogen, methicillin-resistant *Staphylococcus aureus* (MRSA), which can be acquired anywhere. The selective pressure of antibiotic use is clearly increasing their presence.

But as terrible as these resistant pathogens are, the loss of diversity within our microbiome is far more pernicious. Its loss changes development itself, affecting our metabolism, immunity, and cognition.

I have called this process the "disappearing microbiota." It's a funny term that does not immediately roll off your tongue, but I believe it is correct. For a number of reasons, we are losing our ancient microbes. This quandary is the central theme of this book. The loss of microbial diversity on and within our bodies is exacting a terrible price. I predict it will be worse in the future. Just as the internal combustion engine, the splitting of the atom, and pesticides all have had unanticipated effects, so too does the abuse of antibiotics and other medical or quasi-medical practices (e.g., sanitizer use).

An even worse scenario is headed our way if we don't change our behavior. It is one so bleak, like a blizzard roaring over a frozen landscape, that I call it "antibiotic winter." I don't want the babies of the future to end up like my poor aunts. That is why I am sounding an alarm.

■ ■ ■

My personal journey toward the realization that our friendly microbes are in trouble began on July 9, 1977. I remember the date because it was the first time I heard the name of a microbe, *Campylobacter*, that literally set my life's research into motion. I was a newly minted fellow in infectious diseases at the University of Colorado Medical Center in Denver.

That morning I was asked to see a thirty-three-year-old patient who had come to the hospital a few days earlier. He had been suffering from a high fever and was confused. A spinal tap confirmed that he had meningitis, a serious inflammation of the nervous system. His doctors sent samples of his blood and spinal fluid to the culture lab to determine whether the cause was a bacterial infection and, if so, to find out what kind of bacterium it was. While those tests were pending, they started him on antibiotics anyway because he looked quite ill. They believed that he needed big doses of antibiotics immediately or he would die. They were correct.

The test results revealed a slow-growing bacterium identified as *Campylobacter fetus*, an organism that no one at the hospital had ever heard of. That's why I was called. On the job for nine whole days, I was supposed to know the answers. Yikes

Campylobacter organisms are a genus of spiral-shaped bacteria. Like that of so many tiny corkscrews, their helical shape helps them penetrate the gelatin-like mucus that lines the gastrointestinal tract. But why the odd species name *fetus*? (In biology, each organism is identified first by the name of its genus, in this case *Campylobacter*, and then by its species, in this case *fetus*. Each genus has many species and subspecies. Humans are *Homo sapiens*: of the genus *Homo* and the species *sapiens*.) Digging into the medical literature, I discovered that the microbe had this strange name because it affected pregnant sheep and cattle, causing them to abort. It rarely infected humans. How our patient got infected was a mystery. He was a city man, a musician.

Once we knew the organism, we tailored an appropriate antibiotic treatment, and the patient recovered in a couple of weeks. Meanwhile, I was scheduled to give a talk at a clinical conference and decided I would speak about *Campylobacter*. What could be better than talking about a rare infection that no one knew anything about? My own ignorance as a novice would go undetected.

In reading more about *Campylobacter fetus*, I soon learned it has a cousin, *Campylobacter jejuni*. (The jejunum is part of the small intestine.) The literature, scanty as it was, suggested that people infected by *C. fetus* usually have bloodstream infections, whereas those invaded by *C. jejuni* tend to have diarrheal illnesses. Here were two nearly identical organisms with very different effects on the body. Why would one *Campylobacter* stay trapped in the gut, where it kind of belonged, while the other escaped like a ninja into the bloodstream? I was hooked.

Over the next several years, moving from academia to the Centers for Disease Control and back to academia (University of Colorado and Vanderbilt), I became an expert in *C. fetus*, my "favorite" bacterium, and discovered some secrets about its Houdini-like nature.

In this respect, *C. fetus* played an early role in the evolution of my disappearing microbiome hypothesis by teaching me fundamental lessons about how bacteria can persist in their hosts. Yes, they cause disease but, as I later came to appreciate more fully, there also are bacteria that live in us, using a variety of similar tools to escape our immune system. They usually don't harm us; rather, they protect us. I learned that bacteria employ countless tricks, honed from millions of years of trial and error, to do their business, which might either help or hurt their hosts, depending on the circumstances. I will discuss this concept in depth.

C. fetus in particular taught me about stealth—how microorganisms acquire means of escaping host defenses. While 99.9 percent of all bacteria, including *C. jejuni*, are killed by factors found in blood, *C. fetus* glides into the bloodstream by donning a kind of "cloak of invisibility." Even so, it can be trapped by the cells within a heathy liver. But if it is not cleared from the blood in someone with an injured liver (I later learned that the patient whom I'd seen earlier was a severe alcoholic), meningitis can result.

While I was working on *C. fetus* and *C. jejuni* in the early 1980s, a new relative of *Campylobacter* was discovered in, of all places, the stomach. Dubbed "gastric campylobacter-like organism" or GCLO (we now call it *Helicobacter pylori*), it turned out to possess a bag of tricks that, in Jekyll and Hyde fashion, can do us harm or protect us from harm. I have been chasing this organism for the past twenty-eight years, for I believe, and hope to prove, that it is the bellwether that can help solve the puzzle of our modern plagues.

My first encounter with that organism occurred in October 1983 at the Second International Workshop on Campylobacter Infections in Brussels, where I met Dr. Barry Marshall, a young doctor from Australia, who had discovered GCLO and claimed it caused gastritis and ulcers. No one believed him. Everyone "knew" that ulcers were caused by stress and excess stomach acid. I, too, was skeptical. I could easily see that he had discovered a new bacterium but, to me, he didn't have much evidence then at all about ulcers.

It was only in the next two years, after other scientists had confirmed that there was indeed an association of the microbe with gastritis and ulcer disease, that I decided to see if I could make some contributions to understanding the nature of GCLO (which in 1989 was renamed *Helicobacter pylori* after genetic analysis revealed it to be in a separate genus from *Campylobacter*). Their relationship is a bit like that of lions (*Panthera leo*) and house cats (*Felis catus*): relatives for sure but far enough apart to be in different genera. My lab developed a blood test for the microbe and showed that if you carry it, your body has natural defenses against it.

To their great credit, Marshall and his research partner, Robin Warren, conducted clinical studies that showed that eradicating *H. pylori* with antibiotics cured ulcers. Others confirmed and extended their observations. Marshall and Warren went on to win the 2005 Nobel Prize in Physiology or Medicine for this work.

Meanwhile, physicians everywhere commenced an all-out war against *H. pylori* by prescribing antibiotics to anyone with gastric discomfort. Their mantra ultimately became "the only good *H. pylori* is a dead *H. pylori*." I too had been on this same bandwagon for almost a decade.

But by the mid-1990s, I began to change my mind. Evidence was beginning to suggest that *H. pylori* is a member of our normal gut flora and plays a critical role in our health. It was only when I let go of the dogma proclaiming that "gastritis is bad" that I was able to re-evaluate the biology of *H. pylori.* Yes, *H. pylori* can be very harmful to some adults, but later we found that it may be very beneficial to many of our children. Eliminating it may be causing more harm than good. The details of my transition and the reasons for it are laid out in chapters 9, 10, and 11.

In 2000 I moved to New York University and set up a laboratory dedicated to studying how this ancient bacterium did its business in our stomach and what were the consequences to us. Over the following fourteen years, I have accumulated more and more evidence that the disappearance of this venerable microbe might be contributing to our current epidemics. And *H. pylori* led me to a broader study: the human microbiome itself.

These days my lab is bustling. We are currently working on more than twenty projects, looking at how antibiotics affect resident microbes and their hosts, in both mice and humans. In a typical animal experiment, we give mice antibiotics in their drinking water and compare them to mice that do not get the medication. We start very early in life, sometimes just before birth, and then we let the mice grow, studying how fat they become, how their livers are working, how immunity is developing in their gut, how their bones are growing, and what happens to their hormones and to their brains.

To us, the work is exciting, because in each of those areas we can see changes induced by early-life exposure to antibiotics. We have realized that early life is a key window of vulnerability. Young children have critical periods for their growth, and our experiments are showing that the loss of friendly gut bacteria at this early stage of development is driving obesity, at least in mice. We are just beginning studies on social development and celiac disease. We have many ideas for how we can apply our findings from mouse studies to humans. Ultimately, we seek to reverse the damage seen in people around the world, including establishing strategies for restoring the missing microbes. A key step in all of our approaches is to reduce the overuse of antibiotics in our children, starting now.

My odyssey over the nearly thirty-seven years since I saw that ill man in his hospital bed shivering with fever has convinced me that I am at the most critical stage of my own career. The years of working as a doctor specializing in infectious diseases and conducting scientific experiments have given me important perspectives about our modern plagues. I did not anticipate this direction when I was just starting out. But, like a series of transports, the work has carried me across the plains, mountains, and oceans of scientific medical research. It has led to new concepts about our changing modern life that I now want to share with you. The plagues today differ from those that affected my father's sisters, but they are deadly as well.

2.

OUR MICROBIAL PLANET

Our planet began as a lifeless sphere of molten rock about 4.5 billion years ago. But a billion years later, its oceans were teeming with free-living cells. Somehow, in ways still opaque to science, life arose in those primordial seas. Some say the first building blocks arrived as dust falling from outer space, the so-called panspermia hypothesis. Others argue that self-replicating molecules coalesced in clay deposits at the bottom of the ocean, in hot hydrothermal vents or in foamy bubbles created from waves breaking on the shores. We still don't have an explanation for how it all began.

Nevertheless, we have some pretty basic ideas of how life operates, how simple rules give rise to complexity, and how the richness of our planet's diversity came into being. All of biology—the ratcheting of life—rests on the enduring principles of evolution, competition, and cooperation first forged in those oceans.

We live on a microbial planet that is totally dominated by forms of life too small to be seen by the naked eye. For about 3 billion years, bacteria were the sole living inhabitants on Earth. They occupied every tranche of land, air, and water, driving chemical reactions that created the biosphere, and set conditions for the evolution of multicellular life. They made the oxygen we breathe, the soils we till, the food webs that support our oceans. Slowly, inexorably, through trial and error across the deepness of time, they invented the complex and robust feedback systems that to this day support all life on Earth.

It's difficult for the human mind to grasp the concept of deep time, of billions of years of microbial activity churning inorganic matter into the stuff of life. It is a concept emanating from geology, from our understanding of how continents formed, drifted, broke apart, crashed into one another, built mountain ranges and eroded them away with wind and rain over billions of years. And yet the bacteria were here long before even the giant supercontinents, Laurasia and Gondwana, which go back a half billion years or so and from which our present continents have descended.

John McPhee in his classic book *Basin and Range*, collected in *Annals of the Former World*, captures our place in this vast chronology with a wonderful analogy: "Consider the earth's history as the old measure of the English yard, the distance from the king's nose to the tip of his outstretched hand. One stroke of a nail file on his middle finger erases human history."

Or consider another comparison. If 3.7 billion years of life on Earth were compressed into a twenty-four-hour clock, our hominid ancestors would have appeared 47–96 seconds before midnight. Our species, *Homo sapiens*, arrived on the scene 2 seconds before midnight.

But there is another mind-boggling concept needed to appreciate the vastness of our microbial world. Microbes are invisible to our naked eye, with a few exceptions that reinforce the rule. Millions can fit into the eye of a needle. But if you were to gather them all up, not only would they outnumber all the mice, whales, humans, birds, insects, worms, and trees combined—indeed all of the visible life-forms we are familiar with on Earth—they would outweigh them as well. Think about that for a moment. Invisible microbes comprise the sheer bulk of the Earth's biomass, more than the mammals and reptiles, all the fish in the sea, the forests.

Without microbes, we could not eat or breathe. Without us, nearly all microbes would do just fine.

The term *microbe* refers to several types of organisms. In this book, I will be talking mostly about the domain of bacteria, also called *prokaryotes*, single-cell organisms that lack a nucleus. But that doesn't mean they are primitive. Bacterial cells are complete, self-contained beings: they can breathe, move, eat, eliminate wastes, defend against enemies, and, most important, reproduce. They come in all shapes and sizes. Some look like balls, carrots, boomerangs, commas, snakes, bricks, even tripods. All are exquisitely adapted for how they make a living in this world, including those, as I'll elaborate in the next chapter, that thrive on and within our bodies. When they go AWOL, we are in trouble.

Another microbial domain, called *archaea*, superficially resemble bacteria but, as their name suggests, they are a very old, very deep branch of the tree of life with different genetics and biochemistry and an independent evolutionary history. Originally found in extreme environments, such as hot springs and salt lakes, archaea actually may be found in many niches, including the human gut and belly button.

The third branch of microbial life is composed of *eukaryotes*, single cells with a nucleus and other organelles that provide the building blocks for more complex, multicellular forms of life. Over the last 600 million years, eukaryotes have given rise to insects, fish, plants, amphibians, reptiles, birds, mammals—all the "big" life from ants to redwoods that you can see around you. However, some primitive eukaryotes are lumped in with microbes, including fungi, primitive algae, some amoebae, and slime molds.

Here is another kind of scale. Everyone is familiar with a family tree. Your ancestors are lined up by generation with the oldest great-grandparent first, followed by your grandparents, and so forth, expanding the numbers with each generation. Now imagine a family tree of all life on Earth. There are so many forms of life that rather than a tree, it looks more like a bush, with branches extending in all directions. Imagine for the moment that it is a round bush, with the first generation, the origin, near the center and the branches extending outward. Next, let's place us humans on the bush, somewhere around eight o'clock on a watch dial.

Now for the quiz. Where is the life-form on farms that we call corn on that bush? All things equal, we don't think that we are so close to corn, which, after all, is a green plant; maybe it is halfway around the bush? Wrong, it is at about 8:01. If humans and corn are so close, who is taking up the rest of the bush of life and its branches? The answer: It is mostly bacteria. For example, the distance between *E. coli* and *Clostridium*—two common bacteria—is much greater than the distance between corn and us. Humanity is just a speck in the massively bacterial world. We need to get used to that idea.

And then there are viruses, which are, strictly speaking, not alive; they propagate by invading and co-opting living cells. We think about viruses like the flu, the common cold, herpes, and HIV as problems of humans. But most of the viruses in the world are completely irrelevant to us; their targets are bacterial cells, not animal cells like ours. In the oceans, the number of virus particles is unfathomable, more than all the stars in the universe, living off the myriad bacteria in the waters. Over the billions of years that viruses and microbes have been duking it out, each has evolved weaponry to defeat the other. It reminds me of the classic *Spy vs. Spy* comic strip in *Mad* magazine. In fact, one possible treatment for bacterial diseases in humans involves harnessing *phages*—viruses that kill bacteria—an idea I discuss near the end of the book.

While many types of microbes inhabit and shape our world, my main focus here is on bacteria and what happens when we kill them indiscriminately with potent drugs. Although there are plenty of eukaryotes (such as *Plasmodium falciparum*, one of the major causes of malaria) that lead to great misery, the problems they pose are of a different nature. And there are viruses that cause much harm—think about HIV—but they do not respond to antibiotics and are a topic for a different day.

■ ■ ■

Microbes make their homes everywhere we look. The ocean is home to unimaginable numbers of them, though some estimates give a flavor to their ubiquity. At least 20 million types of marine microbes (possibly a billion) make up 50–90 percent of the ocean's biomass. The number of microbial cells in the water column, meaning sea surface to sea floor, is more than 10 to the 30th, a nonillion, or 1,000 × 1 billion × 1 billion × 1 billion. This is equal to the weight of 240 billion African elephants.

The International Census of Marine Microbes, a decade-long project that has been sampling marine microbes from more than twelve hundred sites around the world, estimates there may be one hundred times as many microbial families (genera) as previously thought. Everywhere scientists have looked, some species dominate in terms of numbers and activity. But in what came as a surprise, they also found many species represented by fewer than ten thousand individuals (a puny number for bacteria), including one-off singletons. They concluded that many rare bacteria in the oceans are lying in wait, ready to bloom and become dominant if environmental changes favor them. The same concept holds true for the microbes that inhabit our bodies. The ability to "lurk" for long periods of time in small numbers and then spontaneously "bloom" is an important feature of microbial life.

Many marine microbes are so-called extremophiles. They live in hydrothermal vents where boiling water rich in sulfur, methane, and hydrogen rises from the mantle to meet frigid water, forming conelike chimneys. It is a hellish mixture of acids and heavy chemicals, but it is one in which rich communities of bacteria thrive in the absence of oxygen and sunlight. We see the same thing in the superhot pools and geysers at Yellowstone National Park in Wyoming and in the bubbling tar lake found on the Caribbean island of Trinidad.

Bacteria also live in the massive glaciers of Antarctica and under the frozen depths of the Arctic Ocean.

Oceanic crust composed of dark, volcanic rock at the bottom of the sea, encompassing 60 percent of Earth's surface, is home to perhaps the largest populations of microbes on the planet. Its resident microbes live off energy obtained from chemical reactions between water and rock.

Recently, bacteria have been found munching on plastic particles floating in the open oceans. Although a slow process, at least one thousand different species are involved in converting this "plastisphere" to a healthier biosphere. Other than dump plastic in the ocean, we didn't do anything to stimulate these bacteria. From among the countless varieties floating about, some found their way to the plastic, and those that found it a favorable food source grew in numbers—natural (plastic) selection in action.

The deepest place on Earth, the Marianas Trench, was recently found to support an active microbial community with ten times more bacteria than those in the sediments of the surrounding abyssal plain. And gigantic mats of microbes—the size of Greece—live on the seafloor off the west coast of South America by consuming hydrogen sulfide.

Abundant microbes are lofted by winds, including hurricanes, where they persist and may even make their living high in the skies. They help form cirrus clouds and nucleate ice particles to make it snow. They influence both weather and climate as well as recycle nutrients and decompose pollutants.

Down on the ground, microbes are in charge of soil, one of our most precious resources. Projects to sample soil bacteria worldwide are just getting under way in what some experts call the search for Earth's dark matter, an undertaking akin to figuring out the nature of unknown realms of the cosmos.

We know that microbes make the planet habitable. They decompose the dead, which is a very useful service. And they convert or "fix" inert nitrogen in the atmosphere into a form of free nitrogen that can be used by living cells, benefiting all plants and animals. After the Deep Water Horizon oil spill in the Gulf of Mexico, bacteria ate up much of the contamination because they were able to supplement the nutrients in the oil with nitrogen that they could remove from the air to form a complete meal for themselves.

Microbes live in rocks. For example, in South Africa's Mponeng gold mine, bacteria survive with the help of radioactive decay as uranium splits water molecules, releasing free hydrogen, which the bacteria combine with sulfate ions to make dinner. They even mine the gold. *Delftia acidovorans* uses a special protein to convert floating ions of gold, which are toxic to it, into an inert form of the metal that precipitates from the surrounding water and accumulates in mineral gold deposits. Meanwhile, perhaps the world's toughest bacterium, *Deinococcus radiodurans*, lives on radioactive waste.

But my favorite was described several years ago. Geologists were drilling an exploratory well and studying the cores that came up. From one core taken a mile down, they found only three constituents: basalt (a form of bedrock), water, and bacteria—loads of them. These bacteria made their living and reproduced on just rock and water.

Finally, whole industries are based on harnessing microbes to do our bidding, from making the bread that nourishes us, the alcohol we drink, to the modern drugs engineered by the biotech field. It is fair to assume that bacteria can do just about any chemical process that we might assign to them. In their endless variety are found untold capabilities. We just have to define the problem and go after the right microbes to solve it or we will need to reengineer them. But those exciting possibilities are subjects for another time.

■ ■ ■

The story of microbes is a saga of limitless warfare and also endless cooperation. Since most people are familiar with Darwinian competition and survival of the fittest, I'll start there.

Darwin's careful observations showed that there always was variation among individuals of any species, from birds to humans. He developed his theory of evolution by positing that when variants exist, nature will "select" the one(s) that are best adapted ("the fittest"); these are the ones that best complete their life cycle and have descendants. They outcompete the other variants. Over the course of time, they will crowd out their competitors, even to the point of extinction. It is this natural selection that leads to the commonly stated "survival of the fittest." But Darwin did not know it also pertained to microbes. Like us, he was focused on things he could see—plants and animals— but the fact is that some of the best evidence for natural selection comes from observations and experiments involving microbes.

For example, I can grow a culture of the common intestinal bacterium *E. coli* by placing a tiny dot of existing cells on a plate that nourishes their growth. After an overnight in a warm incubator, the fast-growing *E. coli* might expand to 10 billion cells. The entire plate is covered with a lawn of *E. coli* cells, the growth so dense that individual colonies cannot be seen. Now let's say that I do the same inoculation to another plate, but I add streptomycin, an antibiotic that kills most *E. coli* strains. The next morning when I look at that plate, I only see 10 isolated colonies instead of the lawn of 10 billion cells. Each little colony, the size of a smallish pimple, might have a million *E. coli* cells. Each colony was derived from a single cell that survived the antibiotic and then multiplied on the plate. How do we explain the difference in outcome when we inoculate bacteria onto plates with and without the streptomycin?

First, we can see that the antibiotic worked. Instead of 10 billion cells on the plate, there only are 10 million, a thousandfold reduction. One way to look at it is that the antibiotic killed 99.9 percent of the cells, allowing only a small number to survive. We also can see that the antibiotic failed to some degree. Some cells survived its actions. Why did these cells survive whereas the others did not? Dumb luck? The answer is both yes and no.

The lucky part is that cells resistant to streptomycin possess a variant of a gene that all *E. coli* need to make proteins essential for their survival. The variant gene is not particularly efficient, but it's good enough to help the resistant strains survive and keep making descendants. The susceptible cells die because the antibiotic interferes with the usual version of the same protein, which is essential for cell growth.

These genetic variants that confer resistance arise in an interesting way. It's possible that a few of the cells (ten to be exact in this illustration) in the original culture of a billion cells had the variant gene. They were preexisting. To put these experiments into Darwinian terms, the presence of streptomycin "selects" for the variants in the population that have a resistant form of the gene, whereas the absence of streptomycin in the environment "selects" for the usually more efficient streptomycin-susceptible form. The frequency of the *E. coli* cells with streptomycin resistance would depend on how often streptomycin was around and also how recently. This is a simple example of natural selection, but competition is eternal. May the best microbe win.

While bacteria compete with, prey on, or exploit others, we also see countless instances of cooperation and synergy. For example, if a *Bacteroides* bacterium in the gut can detoxify a chemical in the environment that inhibits *E. coli*, then *E. coli* benefits. A one-way helpful relationship like this is called *commensal*.

Interactions are even more powerful when the benefits are mutual. Imagine that *E. coli*'s main waste product turns out to be a good food source for *Bacteroides*. In this case, the two species will tend to congregate in the same environments. Each is doing nothing more than following its own program, but ultimately they help each other; this is *symbiosis*.

Under other conditions, many different bacteria help each other. Perhaps in a fast-moving stream, bacterium A eats the waste products of bacterium B and also sticks to the edges of rocks. Meanwhile bacterium C, which cannot stick, can adhere to bacterium A to avoid being swept away and helps anchor A in place. And B produces a compound nutritious to C. Now we have a situation where bacteria A, B, and C tend to cluster together, to the mutual benefit of all three.

Over the more than 4 billion years of bacterial evolution, with bacteria dividing and new cells coming along as often as every twelve minutes, and astronomical numbers of individual bacteria, there has been nearly infinite variation. From this endless process, individual bacteria have arisen that have populated all available niches on Earth.

Sometimes bacteria can live stably together, forming a consortium. These cooperative groups abound in the environment—in soil, in streams, on decaying logs, in hot springs—nearly everywhere there is life. The earliest unequivocal proof of ancient life is the existence of 3.5-billion-year-old fossilized "microbial mats" found in Australia, consortia that arranged themselves into large, layered sheets forming whole miniature ecosystems. In all likelihood, some layers performed photosynthesis, some breathed oxygen, some performed fermentation, and some ate unusual inorganic compounds. One species' meat is another species' poison; by settling into layers and combining their abilities, their concerted efforts lead to the benefit of all. There are microbes that can form gelatin-like layers surrounding themselves. These thick gels are called *biofilms*. Their composition varies, but biofilms can protect the bacteria from drying out, or from excessive heat, or from the onslaught of immunity. The presence of biofilms helps explain bacterial persistence in harsh circumstances.

Microbes also form consortia and vast webs of cooperative functions not only in soils, oceans, and rocky surfaces but in animals as well. Such organisms in the human body are the central characters in my tale of "missing microbes." The great biologist Stephen Jay Gould provided a frame of reference for all of terrestrial biology when he wrote: ". . . we live in the Age of Bacteria (as it was in the beginning, is now, and ever shall be, until the world ends) . . ." This is the context for human life, background and foreground.

3.

THE HUMAN MICROBIOME

Think for a moment about your vital organs. Your heart, brain, lungs, kidneys, and liver are complex structures that carry out essential functions that keep you alive. Every moment of the day and night they pump fluids, ferry wastes, deliver air and nourishment, and carry the signals that allow each of us to sense and move about the world. When any one of those organs fails, from disease or trauma, we die. It's that simple.

But what if I were to tell you that you have another vital "organ" that helps keep you alive but that you have never seen. This organ is invisible. It is all over you, especially inside you, yet only recently have we started to appreciate the critical role it plays in keeping you healthy.

Perhaps most remarkable about this part of your body is that it seems completely alien. It does not derive from your obviously human cell lines, according to the blueprint of your human genes. Rather it's composed of trillions of tiny life-forms, the microbes and their relatives that you just read about. Although you might think it a stretch to call this assemblage of microbes a vital organ, functionally the microbiome is just that. Unlike your heart and brain, its development begins not in the embryo but immediately at the moment of birth. It continues to develop in the first few years of life by acquiring ever more microbes from the people around you. But don't be fooled. Losing your entire microbiome outright would be nearly as bad as losing your liver or kidneys. Unless you lived in a bubble, you would not last long at all.

The microbes living in you are not a random mix of all the species present on Earth. Rather, every creature has coevolved with its own collection of microbes that carry out many metabolic and protective functions. In other words, they work for us. There is a starfish microbiome and a shark microbiome, even a sponge microbiome. Reptiles such as lizards, snakes, and Komodo dragons each have unique microbiomes. Every owl, pigeon, and bowerbird has its own set of "bugs" devoted to its species. When the species survive, they survive. Mammals, too, from tiny lemurs to dolphins, dogs, and humans, are full of microorganisms specialized for keeping each of them alive and well.

These microbes provide the animals they inhabit with essential services. They are symbionts that help their hosts in exchange for being housed and fed. Termites can only digest wood because of the bacteria living in their guts. Cows absorb nutrients from the grasses they eat thanks to the microbes living in their four stomachs. Aphids, small insects that live on plants, have resident microbes, including a group called *Buchnera*, that began to live inside them more than 150 million years ago. They possess the key metabolic genes that enable making proteins, a trait that allows the aphids to use the sugar-rich sap from plants as a source of food. In turn, aphids provide a good home for *Buchnera*. It's a win-win. Scientists have constructed the evolutionary family trees of *Buchnera* as well as that of aphids. When we compare the structures of the two trees, they are nearly identical. The probability that this could have occurred by chance is infinitesimal. The only answer is that they coevolved:

aphids and their resident bacteria have reciprocally affected one another's evolution for more than one hundred million years.

A close inspection of the mammalian microbiome reveals that, just as your genes for making red blood cells and proteins can be compared to similar genes in other mammals, your microbes also are part of a larger family tree. In that sense, microbial composition can be considered a marker of ancestry and helps explain why you are more apelike than cowlike. This raises an interesting question. Are you and I more apelike because of our mammalian genes or because of our microbial genes? We always assumed it to be the former, but maybe it is the latter. More likely, it is the sum of the two.

As described, your body is an ecosystem much like a coral reef or a tropical jungle, a complex organization composed of interacting life-forms. As with all ecosystems, diversity is critical. In a jungle, diversity means all the different types of trees, vines, bushes, flowering plants, ferns, algae, birds, reptiles, amphibians, mammals, insects, fungi, and worms. High diversity affords protection to all species within the eco-system because their interactions create robust webs for capturing and circulating resources. Loss of diversity leads to disease or to collapse of the system when keystone species—ones that exert a disproportion-ately large effect on the environment relative to their abundance—are lost.

For example, when wolves were removed from Yellowstone National Park seventy years ago, the elk population exploded. Sud-denly it was safe for elk to browse on, and ultimately denude, the tasty willows that line most riverbanks. Songbirds and beavers that depended on willows to nest and build dams dwindled in number. As rivers eroded, waterfowl left the region. With no wolf-kill carcasses to scav-enge, ravens, eagles, magpies, and bears declined. More elk led to fewer bison due to competition for food. Coyotes came back to the park and ate the mice that many birds and badgers relied on. And so on, down a dense web of interactions perturbed when a keystone species was removed. This concept holds in the natural world as well as your microbiome, where the story of the disappearance of the stomach bac-

terium *Helicobacter pylori* that has colonized humans since time immemorial provides a cautionary tale.

■ ■ ■

Your body is composed of an estimated 30 trillion human cells, but
it is host to more than 100 trillion bacterial and fungal cells, the
friendly microbes that coevolved with our species. Think about that:
right now in your body bacterial cells substantially outnumber your
own human cells. Seventy to ninety percent of all cells in your body
are nonhuman. They reside on every inch of your skin, in your
mouth, nose, and ears, in your esophagus, stomach, and especially
your gut. Women have a rich assortment of bacteria in the vagina.

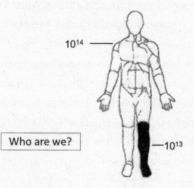

10^{14}

Who are we?

10^{13}

Of fifty known phyla of bacteria in the world, eight to twelve have
been found in humans. But six of them, including Bacteroidetes and
Firmicutes, account for 99.9 percent of the bacterial cells in your body.
The most successful microbes—the winners when it comes to living
with us humans—which descend from only these few lineages, comprise
the basis of a core human microbiome. Over time, they have evolved specialized properties that allow them to thrive in and on particular niches
in the human body. Such traits include the ability to survive acidity,
exploit certain foods, and to prefer dry over wet conditions, or vice versa.

Collectively these bacteria weigh about three pounds, or the same
as your brain, and represent perhaps ten thousand distinct species.
No zoo in the United States contains more than one thousand

species. The invisible zoo living on and inside you is far more diverse and complex.

When you were in your mother's womb, you had no bacteria. But during the birth process and its aftermath, you were colonized by trillions of microbes. Later, we will consider this amazing process. Microbes go from zero to trillions in a short time. There is a well-choreographed succession from the founders to the later inhabitants over the first three years of life.

Ultimately, a unique population of residential microbes develops at each location on the inner and outer surfaces of your body. The crook of your elbow and the spaces between your toes are home to different species. The bacteria, fungi, and viruses on your arms are different from those in your mouth and in your colon.

Your skin is a huge ecosystem a bit larger than a half sheet of plywood, encompassing about twenty square feet of planes, folds, channels, and crannies. Most of these spaces are tiny, even microscopic. Your smooth skin, when viewed up close, may more closely resemble the surface of the moon, pocked by craters with hills and valleys. Which microbes take up residence on what piece of real estate depends on whether the area is oily like the face, moist like the armpit, or dry like the forearm. Sweat glands and hair follicles have their own microbes. Some of your bacteria eat dead skin, some make moisturizers from the oils secreted by your skin, and others keep harmful bacteria and fungi from invading your body.

As for your nose, researchers recently found the signature of many pathogens (disease-causing microbes) living peacefully in the nasal passage of healthy people. One, *Staphylococcus aureus*, is notorious. It can cause boils, sinusitis, food poisoning, and bloodstream infections. But it can also have a completely benign presence in your nose, just minding its own business. At any one time, at least a third of us, and maybe more, are carrying it.

Your intestinal tract is where most microbes in your body make their living, beginning from the top, in your mouth. If you look in the mirror, you can immediately see that there are discrete areas in your mouth, for example, your teeth, your tongue, your cheeks, and your palate. And each site has multiple surfaces. There is the top of your tongue and its bottom. Each tooth has multiple surfaces, and there is a juncture where the tooth descends into the gums. It is fair to say that for every surface there is a different population of bacteria normally living in your mouth. We know a lot about this from the Human Microbiome Project (HMP), a five-year program launched by the National Institutes of Health in 2007. Among the HMP goals was a large project to sequence the genetic material of microbes taken from nearly 250 healthy young adults. One of the take-home messages is that although the overall census of the bacteria present showed a lot of similarities among group members, everyone was unique. Our microbial differences far surpass the differences in our human genes. Our microbes are very personal, a reality we will come back to again and again. Still, there are general principles of organization. We can consider them in the gastrointestinal tract.

In the HMP, the mouth was extensively sampled. Certain families of organisms were found to be common in many sites, such as the Veillonellas, Streptococci, and Porphyromonads, but their distribution varied widely. And other organisms were present only in a limited area.

The richest zone in the mouth is the *gingival crevice*, the interface between tooth and gum. It is teeming with bacteria, many of which are anaerobic (they don't like oxygen). They may be killed by it. It seems counterintuitive that we harbor a big population of oxygen-sensitive bacteria in our mouths, where oxygen-containing air is constantly passing, but it is true. This immediately tells us that there are special niches, some very small, where anaerobic bacteria may flourish.

Ever wonder why your breath smells different in the morning when you wake up? It's because most of time when you sleep, you breathe through your nose. The air exchange in your mouth goes down, and the populations of anaerobic bacteria go up. They produce the chemicals, often volatile, that cause "morning mouth." When you brush your teeth, you are removing tiny debris and whole populations of bacteria. Total counts go down, and the census distributions change. This cycle continues throughout the course of the day.

It is not just in your mouth that microbes cause odor. It is everywhere there are microbes, which in fact is everywhere. In some places, armpits and groin, for instance, microbial concentrations are very high, and the populations are dominated by microbes that produce particularly smelly products. Although whole industries have arisen to control these odors, they are not accidental. From insects on up, our microbial odors identify who we are. They indicate who are our friends, our kin, our enemies, our lovers, or potential mates, and they tell us when might be a good time to mate. Mothers know the smell of their babies and vice versa. Smell is important, and it is mostly microbial in origin. It even determines who is attractive to mosquitoes. Once we fully understand how this works, we might harness that information to become more invisible or repulsive to those pests. But I have digressed.

After food leaves your mouth—where your teeth, saliva, enzymes, and friendly bacteria begin to break it up—it passes into your esophagus, a long tube that separates your mouth and pharynx from your stomach. No one thought bacteria resided in the esophagus until 2004, when we found a rich microbial community of dozens of species living there.

Food then passes into your stomach, where digestion begins with the help of acid and digestive enzymes. Despite strong acidity, bacteria also live there, including *H. pylori* (mentioned earlier), which when present usually dominates. Other species may be found in lower abundance. Your stomach makes hormones as if it were a gland like the thyroid. Its wall contains immune cells that help fight infection, just like your spleen or lymph nodes and colon. *H. pylori* plays a role in the production of acid and hormones and the state of immunity.

Next stop, your small intestine, a long tube that contains the major elements—detergents, enzymes, transporters—for breaking down and absorbing food into your body. This is where you digest most of your meals. Bacteria are present there, too, although in relatively small numbers, perhaps because high levels of microbial activity could inter-fere with the critical functions of nutrient digestion and absorption.

Eventually what remains of the food reaches your colon, where it finds wall-to-wall bacteria. Far and away, most of the microbes in your body live there. The numbers are astounding. One milliliter (about a thousandth of a quart) of colonic contents (and you have several thousand milliliters) contains many more bacteria than there are people on Earth. Your colon contains a universe of bacteria, densely packed, chemically active, accompanying you in your every-day journeys through life. You might think of this as part of the essential bargain of life: we provide them with room and board, and they help keep us alive. But that simplification is not entirely true. Many thousands of people have lost their colon and all of its bacteria because of illness or injury, yet they can live healthy lives for decades. So while this ocean of bacteria that you carry in your colon is very useful, it is not essential. (As mentioned, the same cannot be said for your complete microbiome; its total loss likely would be catastrophic.)

The microbes in your colon break down fibers and digest starch. In one sense, everything that has passed through to the end of your small intestine is on its way out, indigestible by you. But those hun-gry bacteria in your colon can metabolize quite a lot. They can digest the fibers in an apple that has passed through your small intestine and turn them into food—primarily to feed themselves—but some of their products, especially molecules called short-chain fatty acids, are released and actually feed you, starting with the cells in the wall of your colon. They nourish you, their innkeeper.

Up to 15 percent of the calories present in your food are extracted by the guest bacteria in your colon and used to feed you. Like all our resident microbes, they are more than casual or random guests; we coevolved to help each other. Among all mammals, even ones that sepa-rated from one another tens of millions of years ago, there are remark-able similarities in the types of colonic bacteria and in their functions.

The gut environment is warm, wet, and oozy, with numerous dif-ferent neighborhoods occupied by specialized microbes. Some that make particular vitamins might live in particular niches, whereas ones that turn starches into simple sugars may live in much larger neighborhoods. There is competition. As in cities, prized parking spaces and spots in private schools are desirable. Many bacteria hungry for the same nutrients are equipped with identical enzymes and, like lions and cheetahs stalking the same prey, compete vigorously for similar foods. It seems to me that many want to lay their heads on the same soft layers of mucus and use the same limited number of hiding spaces protected from the harsh rain of stomach acid or bile. Meanwhile, many cells lining your gastrointestinal tract are sloughed off every day, so today's hiding place may be tomorrow's sinking ship. By the end, when the last products of digestion leave your body as feces, a mixture of bacterial cells is swept away along with the worn-out cells of your intestinal tract. Together, they and their fragments and water constitute the bulk of your stool.

To give you a sense of their importance in your metabolism, consider that nearly all of the chemicals present in your bloodstream are derived from the activities of your microbes. Bacteria also digest lactose, make amino acids, and break down the fibers in strawberries or, if you eat sushi, the fibers in seaweed.

Through their products, your microbes help you maintain stable blood pressure via specialized receptors located in your blood vessels (oddly, also found in your nose). These sensors detect small molecules created by the microbes that line your intestine. Responding to these molecules affects blood pressure. Thus, after eating, your blood pressure may go down. Could we one day have better treatments for high blood pressure by harnessing these bacteria? Very possibly.

Bacteria metabolize drugs. For example, millions of people around the world take digoxin, derived from the foxglove plant, to treat various heart conditions. How much of the drug reaches the bloodstream depends on the composition of each person's microbiome; the gut is where digoxin undergoes its first chemical processing and then absorption. Variations in the chemistry have consequences. If levels are too low, the drug does not work. If levels are too high, a patient can experience additional heart problems, changes in color vision, and upset stomach. In the future, doctors may be able to gain control over how much digoxin reaches the blood by taming or augmenting gut microbes.

Some of your bacteria make vitamin K, which is necessary for your blood to clot but which is not made by your own cells. It may have been more efficient for the human body to rely on bacteria to produce vitamin K than to go through all of the metabolic costs in manufacturing it ourselves. So our ancestors who acquired vitamin K–producing bacteria were selected over cousins who had to invest in either making it or harvesting a substantial amount from plants. In a sense, our forebears outsourced a key metabolic function to our bacteria. We feed them and house them; they help clot our blood—a wonderful trade.

Some of your microbes even make an endogenous "Valium." People dying of liver cancer often fall into a coma. But if they are given an agent that inhibits benzodiazapines (such as the drug Valium), they wake up. This is because a healthy liver breaks down a natural form of Valium made by microbes in the gut, but a sick liver does not, and the homegrown Valium goes straight to the brain and puts the person to sleep. Other microbes known to live in New Guinea highlanders allow their hosts to live on a diet that is 90 percent sweet potato, which is low in protein. Like bacteria that thrive on the roots of legumes, gut microbes in these New Guinea tribes are able to make proteins from sweet potatoes. They convert or "fix" atmospheric nitrogen found in the highlanders' guts to make amino acids.

■ ■ ■

In women, bacteria colonize and protect the vagina. Until recently medical scientists believed that only one group of bacteria, called lactobacilli, safeguarded the vagina in women of reproductive age from pathogens such as those that cause yeast infections. Indeed, lactobacilli shield the vagina by producing lactic acid, which lowers the pH of the vagina, making it slightly acidic and less hospitable to pathogens. It was assumed that those women whose vaginas are populated by different bacteria would be more prone to vaginal disorders. But now that DNA sequences of the vaginal bacteria from hundreds of healthy women are available, we know that there are five major types of vaginal microbiota, only four of which are dominated by a particular *Lactobacillus*. The fifth type essentially lacks *Lactobacillus*. A woman within this type has several other co-dominant bacterial species in her vagina, but contrary to long-held beliefs this does not make her more likely to develop vaginal disorders, and she is not part of a small minority. About a third of all women have this so-called abnormal mix of vaginal microbes.

Women without lactobacilli have a slightly higher vaginal pH, but their bacteria are just as good as lactobacilli at creating an environment unfriendly to intruders. This kind of functional substitution is probably occurring at sites all over the body, with different bacteria getting the same jobs done in different people.

In addition, we have learned that the bacterial populations in each woman's vagina shift over time. For example, the bacterium *L. inners* may dominate during most of the month, but when a woman has her period, another bacterium, *L. gasseri*, will bloom, only to recede when her menses end. Seems straightforward enough, but this sort of pattern is an anomaly. The most common pattern is that there is no obvious pattern. Sometimes bacteria shift dominance in the middle of a woman's cycle and the next month late in the cycle. Sometimes there are no changes. At other times *Lactobacillus* species take turns dominating the vagina in leapfrog fashion. And in some cases the "abnormal" bacteria dominate, only to disappear without apparent cause. We are still untangling the mystery of what drives these dramatic changes.

■ ■ ■

Probably the most important service your microbes provide is immunity.

In fact, your microbes constitute an important third arm of the immune system. First, there is innate immunity, based on the fact that most of the microbes with which we are in contact have structural patterns that are "seen" by proteins and cells that guard our surfaces. Then adaptive immunity is based on the recognition of highly specific chemical structures. And microbial immunity is based on the microbes that are already in your body, your long-term residents, inhibiting outsiders through various mechanisms. We'll explore each of these in more detail in coming chapters.

Interactions between the immune system and microbes begin at birth, shaping one another throughout your life. It makes sense. One essential property of your resident organisms is that they resist invaders. In essence, your friendly bugs are happy where they live and with the living they make. They do not want outsiders coming in. For example, when invaders try to gain a foothold in the intestines, they must first pass the gauntlet of your stomach acid, which is designed to kill most bacteria; the acid comes from the host, but its production is stimulated by resident bacteria, like *H. pylori*. If an outsider does reach your gut, it must find a source of food, a place to settle. But it's crowded down there. Your resident bacteria don't want to give up their hard-earned spots clinging to your intestinal walls. They certainly don't want to share their meal. So they secrete substances, including their own antibiotics, which are poisonous to other bacteria.

Some invading microbes may gain a toehold for a few days and then be gone, a scenario that happens much more often than not. The fact is, your microbes keep things pretty stable. When you kiss someone, lots of organisms pass between you. But after a while—minutes, hours, days at most—you and your partner will look like you did, in terms of your microbes, before the kiss. There are exceptions (you can acquire harmful pathogens from your lover), and I will get to them. But your ability to resist invaders, even from someone attractive enough to kiss, normally is profound. The same goes for sexual intercourse. There is an exchange not just of fluids but of microbes, and there are changes in both hosts. But after a while, you and your lover are back to how you were before, like nothing (microbially speaking) ever happened. It is possible that some microbes may migrate between partners with regularity, but so far we don't have any candidates, with the exception of pathogens, which often have evolved techniques for spreading among individual hosts.

■ ■ ■

Even changes in diet may not change your microbes all that much. Over time, months and years, the composition of a person's gut microbiome is relatively stable, but yours and mine are different. In one small study, people ate a Mediterranean diet for two weeks: high fiber, whole grains, dry beans/lentils, olive oil, and five servings of fruits and vegetables each day. This diet is strongly associated with a reduced risk of cardiovascular disease. The subjects gave blood samples for the analysis of lipids that have been correlated with heart disease and stool samples to determine which microbes were present before and after dieting. The researchers found a decrease in total cholesterol and a lowering of so-called bad cholesterol, or LDL—a good thing indeed. But the dieters' microbes did not change. Instead each person appeared to have a unique microbial signature, like a fingerprint. The signature remained true, even after manipulation of his or her diet. Yet in other studies of diet, the changes in microbial populations were more significant. In a recent study, changing diet to exclusively plant origin or animal source led to extensive changes, but these lasted only as long as the person was consuming the special diet. We do not know if the diet were to be continued for a year whether the changes would become permanent. We will have to carry out many more studies to better understand the effects of diet on gut microbes. But for now it seems as if relative proportions of the various bacteria in your gut go up and down within discrete boundaries. Research is now aimed at understanding those borders and the extent to which yours and mine are the same and the degree to which they change over a lifetime.

If you are host to 100 trillion microbes and each microbe is a tiny genetic machine, how many genes are cranking away within your resident microbes and what are those genes doing?

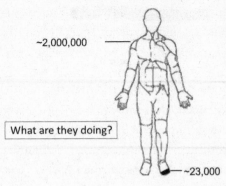

~2,000,000

What are they doing?

~23,000

As discussed above, among the goals of NIH's Human Microbiome Project was to sequence the genetic material of microbes taken from healthy young adults. Not only was a census conducted that defined which microbes were present ("who is there") but also the genes that they carried and their functions ("what is there"). The main findings suggest that your microbes and mine have millions of unique genes, and a more current estimate is 2 million each. Your human genome, by comparison, has about 23,000 genes. In other words, 99 percent of the unique genes in your body are bacterial, and only about 1 percent are human. Our microbes are not mere passengers; they are metabolically active. Their genes are encoding products that benefit them. Their enzymes can produce ammonia or vinegar, carbon dioxide, methane, or hydrogen that other microbes use as sources of food and, in ways we are still working out, they also make many more complex products that benefit us.

A recent survey conducted by a large group of scientists in Europe (begun as the MetaHit consortium) showed something else. A census of nearly three hundred Europeans showed that the number of unique bacterial genes in subjects' guts varied dramatically. The distribution of individuals wasn't normal; it was not a bell curve. Instead, researchers found two major groups. The larger group of 77 percent of the people had an average of about eight hundred thousand genes. The smaller group (23 percent of the subjects) had only about four hundred thousand genes. Two distinct groups; this was not expected. But the most interesting observation is that the people who had the low gene counts were more likely to be obese. This was a striking result, which we will discuss in more depth later.

■ ■ ■

Understanding the ecological structure of our resident microbes presents a tricky puzzle. In a large ecosystem, say a forest, ecologists can directly observe numerous individuals and species behaving and interacting in real time, on daily, seasonal, and annual scales. But we can't yet study microbial ecosystems in anywhere near the same way. As mentioned above, one of our best current methods is to count and identify all the genes in a given community. As a task, that is a bit like scooping up an acre of forest, putting it through a gigantic blender, and then counting the leftover fragments of leaf, wood, bone, roots, feather, and claw, and deducing from the detritus what we can about the woodland's species and their interactions.

We can figure out some functions of our bacterial genes by comparing them with other known genes. Initial findings from the Human Microbiome Project and from the European MetaHit program account largely for what we call "housekeeping" genes, because they are both routine and necessary for life. For example, genes for cell-wall manufacture and maintenance abound since all bacteria have to build cell walls. Similarly, all bacteria must have genes that allow them to replicate their own DNA so they can reproduce. Genes that code for a crucial enzyme, DNA polymerase, needed for making new strands, have been identified. Humans have several varieties of this gene, whereas your resident microbes may have thousands, each one slightly different, depending on which bacterium it comes from.

There also are less subtle differences in the genes of microbes found in different areas of the body. While genes for housekeeping tasks remain consistent, skin bacteria have more genes related to oils than do bacteria living in the colon. Vaginal bacteria have genes to help them create and deal with acidic conditions. At this point in our knowledge, we can safely predict that bacteria will carry out specialized functions in each of the body's habitable niches and that the differences involved are much greater than those seen in the human genome. For example, the difference in height between the tallest and the shortest adult on Earth is perhaps two- or threefold. Organisms in a typical microbiome may range, in their individual representation, by a staggering ten million–fold. Bacterial specialization is a thrilling and largely unexplored realm in uncovering what makes each of us distinct in terms of our health, metabolism, immunity, and even cognition.

While we have yet to identify the function of some 30 to 40 percent of bacterial genes identified by the large projects, we do know that some species are rare and vulnerable to extinction. As with vaginal microbes, bacterial populations can be extremely dynamic. The number of cells representing a particular species can vary from, say, one cell to a trillion. Let's assume an animal is exposed to a new food that contains a chemical never before encountered. The bacterial species that is today represented by one hundred cells could, given a triggering change in the intestinal environment such as the new food, become billions of cells within a few days. If faced with loss of a prized food or with competition by its hungry fellow bacteria, the numerically dominant species could then drop in numbers several thousand fold or more. It is this dynamism and flexibility that are at the heart of the microbiome and contribute to its staying power. But the species represented by a hundred cells in normal times doesn't have a big margin for error. It could also encounter an antibiotic that wipes it out permanently.

I call these rare species contingency microbes. Not only can they exploit an unusual food chemical (which more common bacteria cannot), but they may provide genetic protection against threats, such as a plague that humans have not before encountered. To me, this is a flashing red light. Diversity is essential. What if we lose critical rare species? What if human keystone species disappear? Would there be cascading effects leading to secondary extinctions?

■ ■ ■

The fact that we can coexist with bacteria raises a profound set of questions. Why don't they wipe us out? Why do we tolerate them? In the dog-eat-dog world of Darwinian competition, how have we achieved a stable relationship with our microbes?

Public-goods theory provides clues. A public good is something that everyone shares, such as the clean air you breathe at the seacoast, a bright sunny day, a local street built with your tax dollars, or your favorite public radio station. But nothing is ever really free. Public radio must be supported; someone has to pay. Even if clean air is public, your car might emit pollution that affects my clean air. My breathing and your driving occur in the same space.

In a smoothly functioning social world, each individual is expected to contribute to the public interest. You can listen to public radio and not pony up but, if everyone did that, public radio would go bankrupt. If everyone had a car out of tune, our common air and sunlight would be degraded. In this sense, people who use a public good but don't give sufficiently, or who add to the common expense, may be considered "cheaters"; they benefit but do not pay their fair share of the enterprise.

However, out in the jungle, where "survival of the fittest" rules supreme, "cheating" seems like a pretty good strategy. The cheater might be able to lay more eggs or find better nesting sites and, over generations, be more successful (have more offspring) since its ratio of benefit per cost is more favorable. The cheater has a selective advantage. However, if "cheaters" always won, cooperation would fall apart. Why wouldn't everyone cheat and not pay for public radio? How can different life-forms live together if there is a built-in selective advantage for breaking the rules? Cheating has the power to make the whole system break down.

Yet clearly cooperation occurs everywhere we look: bees and flowers; sharks and pilot fish; cows and their rumen bacteria allowing them to create energy from grass, termites, aphids. As far as we know, ruminants have existed for millions of years and insects like termites and aphids even longer. This tells us that cheaters don't always win. Simply put, the penalty for cheating must be sufficiently high that cheating is disadvantageous, so that cheaters don't triumph. If there were no consequences, more people would speed when they drive. Penalties work.

The same holds true for you and your microbes. Natural selection favors hosts that have a system of penalties in place that cannot be evaded: the more the cheating, the higher the penalty. Such penalties can deflect the spoils of "ill-gotten" gains. Thus a bacterium in the termite gut that oversteps its bounds can trigger a very strong immune response, putting it back in its place. This works, but it can be expensive for the host to have such a system. Some might die fighting off cheaters with an overly aggressive immune response. When the host dies, so do all of its inhabitants. When this happens, all of the genes, from both the host and its residents, are lost for all of posterity. Other termites that did not have a cheater arise and take up the niche vacated by their newly deceased sibling. The tension between competition and cooperation plays out on a thousand stages.

Game theory, inspired by the great economist and mathematician John Nash (whose story has been told in the book and movie *A Beautiful Mind*), sheds light on the phenomenon of cooperation, on why coevolved systems appear to select for individuals who largely play by the rules. It is a way of understanding behavior in social settings—how people make decisions to optimize outcomes and how markets operate. Nash envisioned a situation that has since been called the "Nash equilibrium." It can be summarized as a strategy in a game with two or more players in which the outcome is optimized by playing within the rules; if you cheat, your outcome is worse than if you played fair and square.

Ecosystems that have been around a long time, like our bodies, have solved this fundamental tension between conflict and cooperation. We have persevered. But this theory has relevance as we consider our changing world. What that means is that cooperation is tenuous: don't mess with it, because then all bets will be off. I worry that with the overuse of antibiotics as well as some other now-common practices, such as Cesarian sections, we have entered a danger zone, a no-man's-land between the world of our ancient microbiome and an uncharted modern world.

4.

THE RISE OF PATHOGENS

When I was a medical student, I spent the summer assisting a doctor whose job was to examine workers in a West Virginia Job Corps program. It was a great experience, because it was intensely clinical. I learned to do careful physical exams on a large number of basically healthy young people. My teacher, Dr. Fred Cooley, was practical, smart, and funny. My job with him ended at about one in the afternoon, so I could head over to the hospital and work with other doctors seeing all kinds of patients. They didn't have many medical students, so they welcomed me with open arms, a trainee with lots of questions.

One afternoon, we were called to see an eleven-year-old boy who had become acutely ill and was hospitalized. He lived in a small, very conservative, Baptist community. He had been perfectly well until about two days earlier, when he began to feel achy; he developed a fever and an upset stomach. The next day his fever worsened, and he had a headache. On the third day, he developed small purplish dots all over his body. His parents were scared and brought him to the hospital, which was a good thing. The emergency room doctors quickly diagnosed Rocky Mountain spotted fever, a disease caused by a bite from a tick infected with a type of bacteria called rickettsia. Although first discovered in the Bitterroot Valley in Montana, hence its name, it is much more common in the eastern half of the country.

The bacterium multiplies within cells lining blood vessels, invoking a vigorous immune response. That explains the rash, since the blood vessels get inflamed and break. And it explains the headache, since the brain's blood vessels are involved, causing a form of encephalitis. The boy was started on tetracycline, a life-saving antibiotic. If treatment isn't given immediately, or is started too late, RMSF is fatal in up to 30 percent of people.

I accompanied the doctors up to see him. His hospital room was darkened because light hurt his eyes, indicating that his brain was affected. His body was covered in purple spots, more than I have seen in anyone since. Some of the spots ran together, producing big blackish purple patches. His hair was matted. He was drenched in sweat as he thrashed from side to side, his hands tied to the bed rails so he would not hurt himself and others. He was yelling at the top of his lungs, hallucinating, completely incoherent. From time to time, a recognizable word would emerge, but they were all curses: "shit, fuck, you fucking bastard, tits, cunt, fuck." This went on continuously. In the corner of the room, his parents were cowering. Where had he learned these words? We knew that the encephalitis was causing his lack of inhibition.

Fortunately, with treatment, he turned the corner, gradually got better, and was discharged from the hospital five days later to complete his treatment at home. He didn't remember anything that happened, but I am sure his parents never forgot it, not just the horror of his illness but the miracle of his cure.

Pathogens like rickettsia are microbes that make you ill. They bring the fevers, chills, pains, and aches that keep you bedridden for days. They can kill you—slowly or rapidly. You might die alone or alongside thousands of others. We usually call them germs and have, since their discovery about 150 years ago, done everything in our power to kill them. For the past 70 years we have waged an aggressive war against pathogenic bacteria using a slew of antibiotics, saving millions of lives worldwide. But to our chagrin, this battle seems to have no end. Bacteria mutate with lightning speed and have developed resistance against some of our most effective antibiotics. Even more worrisome, the battle we wage against pathogens has led to serious unintended consequences for our health and well-being.

But before we look at those consequences, let's get to know what we are fighting. Other than potentially harming us, pathogens differ in many ways, for example, their biological nature—are they bacteria or viruses? Do they produce a toxin that injures our cells while they live offshore, in the middle of the gastrointestinal tract, like a battleship lobbing in shells along the coastline? Or are they like the marines, aggressively coming onshore and inflicting damage that way?

It's tempting to think of pathogens as intrinsically evil, but they're not. Just like Yellowstone's wolves, they are predators. More often than not, by pursuing their own survival, pathogens inflict terrific damage on the hosts they inhabit. Sometimes the damage is accidental, the pathogen's cost of doing business. But for pathogens that are well adapted to their host, the damage serves a purpose. For example, the bacteria that cause tuberculosis make people cough, thereby spreading themselves around and infecting other people. Similarly, the rabies virus attacks the part of the brain involved with aggressive biting behavior and is spread via saliva in infected animals.

David Quammen, in his book *Spillover*, about emerging infectious diseases, notes that we think of predators as big beasts that eat their prey from the outside, whereas pathogens are small beasts that eat their prey from within. It's an apt description.

The Inuit believe that "the wolves keep the caribou healthy." A healthy caribou can easily ward off wolves, whereas wolves spot weaker members of the herd, rush in, and tear them apart for dinner. They thin the herd. It's the same with pathogens. Seven billion people live in today's world, often existing in squalid, crowded conditions. Malnourished, weak, and often without access to modern drugs, impoverished humans can be easy prey for the pernicious pathogens discussed in this chapter. I'm not saying that thinning the human herd is a good thing. Just that it has always happened and assuredly will happen again.

There are pathogens that simply get under your skin through cuts and scrapes. When a wound is not properly cleaned, you can get an infection, but it is treatable: if mild, just with cleansing, a band-aid, and a kiss; if more severe, then with deep cleansing. Sometimes antibiotics are needed. These cases are pure accidents. The pathogens almost never spread to another person.

Organisms that ordinarily are not very pathogenic (disease-causing) can evolve extraordinary levels of virulence and can also kill robust, healthy individuals in a very short time. Most of us carry *E. coli* in our intestine, and most strains don't harm us at all. But in 2011 there was a huge outbreak of *E. coli* infections in Germany when people ate contaminated sprouts. At least two *E. coli* strains exchanged genetic material, producing an extremely virulent organism that infected more than four thousand people, damaged the kidneys of more than eight hundred of them, some permanently, and killed fifty.

Communicable diseases are caused by microrganisms that colonize your body, multiply out of control, and make you ill. They can be viruses that cause the flu, bacteria that cause whooping cough, fungi that grow in the lining of your mouth, or a variety of free-living single-cell organisms called protists, such as a nasty amoeba that causes dysentery and bloody diarrhea. More than fourteen hundred human pathogens are currently recognized. They can be high or low grade. The rickettsia that caused spotted fever in the previously healthy young boy is a high-grade pathogen, whereas the kinds of organisms affecting people with chronic lung diseases can be low grade, meaning they are less virulent. They cause illness when a person is compromised and are less likely to make a perfectly healthy person ill.

Ultimately, all communicable disease—causing microbes come to us from our primate cousins, from our domesticated animals, and in

other ways that are increasingly dangerous, including from wild animals. Some "jumped" from animals to humans so far back in the past
that we can't be sure of their origins. But other diseases can be traced:
plague from the fleas that live on rodents, rabies from bats, influenza
from birds, Lyme disease also from rodents but now via ticks. Some
of the deadliest pathogens are rogue viruses that have emerged much
more recently: Ebola, SARS, Hantavirus, Marburg virus, swine and
bird flu. They are virtually impossible to eradicate because we humans
can come into contact with the animals in which they live in all sorts
of ways. When intermediate vectors like mosquitoes help transmit
disease, as with malaria, the picture gets especially complicated.

Some of the most successful human pathogens no longer need
their original animal reservoirs. Somewhere along the line, the smallpox, polio, and measles viruses evolved to specialize in humans; they
affect us exclusively (and thus are vulnerable to elimination once and
for all from humans—like smallpox). But the 800-pound gorilla of
recent pathogens, HIV, which jumped to humans from chimpanzees,
is now transmitted from person to person through sexual intercourse
and contaminated blood. From occasional chance events, over 100
million people are now infected. As I will discuss in chapter 15, I am
concerned that we are creating the conditions favorable for the spread
of other pandemic microbes by the combination of easier global travel
and the lowering of our defenses.

■ ■ ■

For the vast majority of human history and prehistory, the pathogens
behind the world's great epidemic diseases—smallpox, measles, influenza, plague, polio, cholera, typhoid, scarlet fever, and diphtheria—
did us no harm. They did not kill us. The reason has to do with
population size. When our ancestors were hunter-gatherers in central
Africa, they lived in small groups—maybe thirty to sixty individuals—
and were widely dispersed across the vast savanna. They lived this
way for about 2 million years before *Homo sapiens* arose, about two
hundred thousand years ago. Our existence in civilizations going

back about eight or ten thousand years is just a punctuation point on our enormous prehistory. That long period shaped who we are today.

Our ancestors were self-sufficient. When times were plentiful, males brought home enough game to nourish the group; females foraged for fruits, nuts, and plants. But when food was scarce, people suffered. Hunters exhausted themselves trying to find game. Malnourished women stopped menstruating or lactating. Worst of all, when severe droughts persisted, entire groups died out, leaving no trace. Hyenas and vultures picked their bones clean.

But from our modern perspective, this precarious existence had one good thing going for it: there were no epidemics. Our ancestors suffered from common infections, such as parasitic worms and yaws, which are chronic, nonfatal disorders. There were no epidemic diseases because these tiny bands were totally isolated, with no neighbors to bring harmful bacteria or viruses into their communities. If by happenstance a lone individual with a contagious disease stumbled into their settlement, the outcome could go one of several ways: nothing happened, everyone became ill and died, or a few became ill and the rest became immune. But after that, the pathogen had nowhere to go. There were no new hosts to infect. It was marooned and died out.

But the hunter-gatherers did have to contend with latency. Eons ago, tuberculosis and several other well-known pathogens developed this strategy—latency—which permits them to infect one generation, lie low, and then infect later generations, thus avoiding the problem of what happens when no new susceptible hosts are around.

Another example of latency occurs with chickenpox. If, like many children, you breathed in the varicella-zoster virus, you would have soon developed a fever and then broken out in a rash, with blisters all over your body. After a few days, the rash would have faded. In two weeks, you'd be back to normal. With rare exceptions, a child who contracts chickenpox develops lifelong immunity to varicella-zoster. That is the end of the story, or so it might seem. But the virus is clever. It sequesters itself in nerve cells along the spine and in equivalent locations in the head. The virus endures like this for decades, silently, stealthily, causing no discomfort.

Until one day, when you are in your sixties, seventies, or eighties, you might feel a tingling sensation under a rib on one side of your body. The next day, you notice a rash following the contour of your rib. Upon close inspection, you see that the rash has blisters similar to those you had as a child with chickenpox, only this time it is localized. You now have the shingles, which doctors call *herpes zoster.*

In general, the older you are, the more likely you are to get shingles. For decades your immune system is able to hold the virus in check, but as your system weakens with age, the virus is no longer suppressed and out it pops—as zoster. And when the zoster blisters break from their elderly host, the virus spills out into the air, where it can infect a child who has not yet acquired immunity.

And so the cycle repeats. In this way, varicella-zoster can skip entire generations. Although there may be no cases of acute infections in a small community for decades, the virus can "come alive" at any point and then infect a whole new group of susceptible people who were born in the years since the last active transmission. The virus, well adapted to humans, has two opportunities for transmission: from a child with chickenpox or from an aged relative who had chickenpox long before and now has shingles. Contagious, latent, contagious—this is a strategy that optimized success during the long period when our ancestors lived in small bands as hunter-gatherers in the African savannahs.

The bacterium that causes tuberculosis is transmitted similarly, both acutely and after reactivation of a latent infection, usually in an elderly person, which optimized its survival in the small isolated populations that dominated our prehistory. But when human populations later expanded, tuberculosis took off like a rocket.

Small populations are now the exception. About ten thousand years ago, the invention of agriculture made food supplies secure. Populations soared. Trade flourished. Towns grew into cities and crowding was commonplace. And that's when epidemic diseases began to take off.

Measles is the best-known example for illustrating the working of these so-called crowd diseases. Epidemics often occur in waves and spread quickly from one person to the next until nearly everyone is exposed. In a short time, you survive or die. In the case of measles, survivors developed antibodies and remained immune for the rest of their lives.

Caused by the rubeola virus, measles is the most infectious disease known to humankind, with an infection rate greater than 95 percent. In contrast, a new strain of influenza might infect from one-third to one-half of those exposed for the first time. When I worked in Africa as a student, I saw many children with measles. Typically, they had high fevers, inflamed throats, red eyes, and hacking coughs. The cough, which yielded virus-laden aerosols, was very effective in spreading the disease. Any child not previously exposed to the virus became infected immediately. After a week or so of cough and runny nose, a characteristic rash appeared at the back of the ears and then spread to the rest of the body: measles. Now kids across the developed world are vaccinated, but Africa and other less-developed parts of the world have been slow in catching up. In 2011, there were 158,000 measles deaths globally; 432 people, mostly kids, died from measles every day—18 deaths every hour.

For the measles virus to survive, it must encounter a new susceptible person every week or two. Like a Ponzi scheme, it requires new victims with startling regularity. In fact, measles can be sustained only if there is a contiguous human population of 500,000 people. In such circumstances, a 3 percent birth rate provides 15,000 newly susceptible children each year, guaranteeing measles transmission year after year. But we have had contiguous populations of 500,000 for only about 10,000 years, and thus the epidemics they enable. So measles might have jumped from animals to humans many times before in prehistory, but without sufficient population size it died out.

For instance, many islands, like the Faroes in the North Atlantic, were free of measles for decades at a time. But when a ship brought in an infected person, as one did in 1846, the measles virus quickly spread from person to person. Essentially everyone got the disease. A similar outbreak occurred in Hawaii in the mid-eighteenth century, when measles was introduced by a sailor. People burning with fever went down to the ocean to cool off. But it didn't help; when the epidemic was over, one in five people had died. The virus died out, only to return by ship many years later.

The rise of cities brought other dilemmas. We had to store food and that attracted hungry pests and their parasites. Scavengers like rats came to visit our grain bins and trash heaps. Thus arose bubonic plague, transmitted by fleas on rats and caused by the bacterium *Yersinia pestis*. The so-called Black Death erupted in Europe in 1347 and within a decade wiped out a quarter to a third of the population. Once introduced, it could even spread without rats, as infected fleas hopped from person to person, and people with plague pneumonia coughed on others.

In 1993 plague broke out in Kinshasa, Zaire. Years of war and corruption caused the government to print money. As a result, there was hyperinflation. People bought whatever they could today, because tomorrow it would cost more. So they stored a lot of grain. This hopeful act brought rats and the plague they carried into many homes.

The Industrial Revolution caused populations to balloon, and many diseases transmitted from person to person worsened. Scarlet fever caused by streptococcus, diphtheria, typhoid fever, and tuberculosis all ravaged the crowded cities. In 1900, tuberculosis was the leading cause of death in the United States. Diarrheal diseases, spread by the contamination of drinking water with sewage, sickened the growing numbers of susceptible people. Twenty percent of children did not survive to age five because of diarrheal diseases, whooping cough, diphtheria, scarlet fever, and other epidemic diseases.

With larger and larger towns and cities and better connections established through transportation and trade, our indigenous microorganisms—those that were endemic or latent—were increasingly joined by epidemic pathogens that required large contiguous populations to sustain them, and they flourished. These were the real troublemakers, the killers and maimers, especially of children. Even tuberculosis, which had been around for a very long time, evolved strains that were selected for virulence and ease of transmission. Together, all of these pathogens thinned our human herd at enormous cost. Rich or poor, no family was immune. People could only pray for deliverance from pestilence. Not much help arrived until the late nineteenth and early twentieth centuries, when the very first advances in sanitation were made and then followed by the development of vaccines. Through concerted efforts and great international cooperation, vaccines now have eradicated smallpox from the face of the Earth, markedly reduced the reach of polio, and curtailed the measles epidemics. The other incredible advance in the fight against pathogens came when antibiotics were finally, gratefully, discovered.

5.

THE WONDER DRUGS

As I drove to work on a spring morning in 1980, the air in Atlanta held a chill. I had been away for more than two months working in the hot enclaves of Bangladesh and India and was relieved to be back at the Centers for Disease Control. The office held the usual welcomes, big piles of mail, and much to organize, but in the afternoon, I started to feel achy. Maybe it was jet lag; I had arrived the night before. But I felt lousy. My forehead was hot to the touch. After about an hour, I decided to go home. Maybe I had caught the flu on the airplane or on a prolonged stopover in England. I couldn't remember the last time I felt too sick to work. Time to go to bed, and by the morning I would feel better.

But the next morning, I wasn't better. My fever was up to 101°F. As an expert in infectious diseases at the CDC, I knew that malaria can begin like a case of the flu: fever, headache, achiness, sore muscles. Could I have picked up malaria? When travelers die from malaria, it's because the diagnosis is missed and treatment is started too late. People think that they have the flu. With that in mind, I called one of my CDC colleagues in the Parasitic Diseases Branch, Dr. Isabel Guerrero. I wanted to get a blood smear taken to see if it was malaria.

"I'll come right over," she said.

In about thirty minutes, she was at my bedside at home, where she pricked my finger, put a spot of blood on a glass slide, and told me she would call with the results.

About an hour later, she did. "You don't have malaria."

Thus reassured, I was ready to wait out the flu. By then, I had developed a mild cough.

The next morning, Wednesday, I was still sick. I didn't feel too bad, but I still had a fever. My wife convinced me to go see a specialist in infectious diseases, Dr. Carl Perlino. He examined me and, other than the fever that perversely vanished while I was in his office, I checked out fine. Even my screening blood tests were okay.

The next day, Thursday, I still had a fever and the mild cough. I was in bed all day, and that night I had a vivid nightmare. I don't remember who was chasing me, but I woke in a cold sweat. The sheets were drenched. Even in my delirium, I knew instantly what the problem was: typhoid fever! Traveling in Bangladesh and India, where human waste often gets into food . . . symptoms that began about a week after I left . . . day after day of fever, now worsening—vague symptoms. That's what it had to be.

By the next morning, I was very weak. My temperature had shot up to 104°F. I didn't have the strength to button my own shirt or sit up in the car without leaning on the window. I knew that I had about a 10–20 percent chance of dying if I wasn't treated with antibiotics. Achy, sweating, no strength, no intake of food in days but no appetite—I knew that I was acutely ill. As we drove on that exquisite spring day down a street filled with blooming magnolias, I thought that it would be a real shame to die at thirty-one.

When we got back to the doctor's office, I was huddled and shivering. They had to put me in a wheelchair. My greatest fear was that Dr. Perlino would not understand how sick I was and would send me home. It was ironic; I knew that hospitals are dangerous places and should be avoided at all costs—people fall out of bed, they get the wrong medications, they acquire new infections—but I was desperate to be admitted, to start on treatment, not to go home.

Fortunately, he took one look at me and immediately admitted me to the hospital. Another irony is that my main job at CDC was as the *Salmonella* surveillance officer of the United States. Doctors from all over the country would call me to ask advice about patients and outbreaks of salmonella. So here, too, my doctor asked me what antibiotic I should be treated with. I knew that *Salmonella typhi*, the main cause of typhoid fever, could be treated with ampicillin, an advanced form of penicillin. Ampicillin was life-saving for millions of people. But there was a big problem: it had been used so much that by 1980 many strains of *S. typhi* had become resistant. It might be completely ineffective.

So instead I recommended a newer formulation of a sulfa drug, called *co-trimoxazole*. It combined two agents developed in the 1960s and was still widely effective against *S. typhi* (though resistance to it would later develop as well). Evidently, despite my high fever, I could still think straight. Even if I was wrong about typhoid, I was so acutely ill that the doctors had to treat me with something in case I had some other kind of bacteria spreading through my bloodstream.

Medical students came to take samples of my blood to the culture lab. If I had typhoid fever, *Salmonella typhi* would show up in the petri dishes. Then they hung a bag of fluid containing the co-trimoxazole and dripped it into my veins. I knew that the odds were turning in my favor. The chance of dying was getting smaller with each hour. That is the miracle of the antibacterial drugs that started being discovered in the 1930s.

I slept and slept. But the next morning, I wasn't better. Still feeling achy and miserable, I asked the team, "What do the blood cultures show?"

"Nothing growing."

Could my self-diagnosis be wrong? Was it not typhoid? But it was only twelve hours or so since the cultures had been taken, so maybe it was too early. In the odd position of both patient and specialist doctor, I recommended that we continue the course, and they agreed.

The next morning, the team came into my room. "The cultures are positive. You have *Salmonella* in your blood. The microbes are growing."

So it was typhoid after all.

The next day there came a small surprise. It wasn't *Salmonella typhi*, the usual cause of typhoid fever, but *Salmonella paratyphi A*, essentially the twin of *S. typhi*. But according to the textbooks, the cases are indistinguishable, and I could vouch for that.

With treatment, and a few complications, I turned the corner and started to get better every day. After a week, I was discharged from the hospital, and I spent another week at home before returning to work. A week sick at home, a week in the hospital, a week convalescing back at home—this was a serious illness. I shuddered to think what it would have been like without the co-trimoxazole.

A few years later, I was speaking with a colleague who had worked in Asia for many years. I told him that, as far as I knew, my only dietary indiscretion in the weeks before my illness was that one hot night in Mumbai, as I was walking around, I saw a street vendor who was selling slices of watermelon. His stand didn't look that great, so I asked him to give me a slice from an unopened melon. I thought that would protect me. That was about nine days before I started to become ill, which is almost a classic incubation period.

"Of course," my colleague said. "Of course, it was the melon."

"You see," he told me, "in India, they sell the watermelons by weight. So the farmers inject water into them, to make them weigh more. The water comes from the rivers and streams near their fields."

My stomach churned at the thought. The watermelon was contaminated with human waste. You get typhoid fever from ingesting food or water contaminated with the fecal waste from a person who is a carrier of the disease. I thought of the most famous carrier, Mary Mallon, better known as Typhoid Mary, the young Irish immigrant who worked as a cook for well-to-do families near New York around 1900. After an outbreak of typhoid fever in her house, she would move on to another family. And then there would be another outbreak and then another. It is not clear whether or not she understood that she was causing the outbreaks. There was plenty of typhoid around in those days; hospital wards were filled with suffering people, and at that time about a quarter of them died. A great medical detective named George Soper traced the outbreak back to Mary and made her promise to stop working as a cook. She was a typhoid carrier: she felt entirely healthy, and she *was* entirely healthy. Carriers aren't ill; they just shed the organism.

Mary denied that she had anything to do with the prior cases, and within a short time she absconded from her parole. Eventually there was a trail of new outbreaks. Soper found her again. Here was the dilemma: Mary was perfectly well, but a menace to the community, no less than if she fired a loaded gun at random into a crowd. Typhoid was not a mild illness; people died after coming into contact with her cooking. Ultimately, a judge decided. Mary was imprisoned on North Brother Island in New York's East River and lived the rest of her life in custody, swearing her innocence to the end. These days, we could probably cure her condition by removing her gallbladder and giving her antibiotics. And the people she infected could be treated with antibiotics, as I was.

Fast-forward twelve years from Atlanta to May 1992, when I was asked to speak at a conference highlighting advances in our understanding and treatment of infectious diseases. My topic was how our work had linked a newly discovered bacterium in the stomach, *Helicobacter pylori*, to stomach cancer, a common and difficult-to-treat malignancy. It was, we thought, a new pathogen, and people were curious to learn more.

The symposium was held at Yale specifically to mark the fiftieth anniversary of the first use of penicillin in the United States. The moderator began by recounting the case of Anne Miller, a thirty-three-year-old nurse who had suffered a miscarriage in 1942. She had been acutely ill for a month, with fevers up to 107°F, delirium, and signs of a streptococcal infection raging throughout her body. She had childbirth fever, or what doctors called *puerperal sepsis*. It was an infamous killer of young women after a miscarriage or birth of a child. Miller drifted in and out of consciousness, very near death.

By an incredible stroke of luck, her physician gained access to one of the first tiny batches of penicillin, which was not even commercially available yet. The drug was rushed via airplane and state troopers to Yale–New Haven Hospital, where it was administered to Miller on her sickbed.

Her recovery began within hours. The fever broke, the delirium ended, she could eat, and within a month she had recovered. It was the scientific equivalent of a miracle. What made the difference was 5.5 grams of penicillin, about 1 teaspoon worth, diluted into her intravenous solutions. Penicillin was in such short supply that her urine was saved so it could be shipped back to the Merck pharmaceutical company in New Jersey where the excreted penicillin was purified for use in another patient.

As the moderator recounted the details and the drama of the story, the audience was transfixed. You could hear a pin drop. And then, after a short pause, he said, "Now, will the patient please stand up."

I turned around to look. In the third row, a small-boned, elegant, elderly woman with short white hair stood up and, with bright eyes, looked out across the room. She was Anne Miller, then in her eighties, given fifty more years of life by the miracle of penicillin. I can still picture her shy smile. She lived another seven years before her death at age ninety.

When Anne Miller's life was saved, medical science was just beginning to learn how to defeat bacterial infections. Pneumonia, meningitis, abscesses, and infections of the urinary tract, bone, sinuses, eyes, and ears—indeed, all parts of the body—were still being treated with marginal or questionable remedies from the past. When George Washington developed a throat infection, he was bled by a surgeon. Doctors had great faith in this therapy, but it probably hastened the president's demise. Bleeding continued as a remedy into the twentieth century.

Some treatments helped slightly but none dramatically, and the side effects of many patent medicines were worse than the diseases being treated. Some contained high levels of arsenic. Even as surgical techniques improved, infection was a constant worry and could transform a successful operation into a disaster; with bad luck, removing an ingrown toenail could lead to a foot amputation. An infected heart valve was 100 percent fatal, worse than cancer.

During the American Civil War, more soldiers died from typhoid fever and dysentery than from bullets. No one was immune. Leland Stanford Jr., the son of the governor of California and for whom the university is named, died of typhoid fever in Italy. He was fifteen. During World War I, dysentery and typhus took a greater toll than combat. In 1918 and 1919, the Great Spanish Flu spread across the globe to infect 500 million people, about a quarter of the world's population, killing between 20 million and 40 million of them, frequently from complications due to bacterial pneumonia.

Scientists worked feverishly in the late nineteenth and early twentieth centuries to combat infectious diseases. They had one light to guide them: germ theory, the concept that many diseases are caused by the presence and action of microorganisms, especially bacteria.

A handful of brilliant scientists, the giants in their field, led the way. In 1857, French chemist Louis Pasteur showed that fermentation and putrefaction are caused by invisible organisms floating in the air. He demonstrated that meat decay was caused by microbes and that disease could be explained by the multiplication of germs in the body. Following Hungarian physician Ignatz Semmelweis, who markedly reduced deaths due to childbirth fever by requiring hand washing, the British doctor Joseph Lister revolutionized surgical practice by introducing new principles of cleanliness. Inspired by Pasteur, he soaked dressings with carbolic acid (a form of coal tar with antiseptic properties), covered infected wounds, and thereby improved their healing. And Robert Koch, a German doctor, developed methods to assess whether a particular microorganism causes a specific disease; these criteria are known today as Koch's postulates. He also developed stains for visualizing the bacteria that cause tuberculosis and cholera under the microscope.

But while germ theory led to improved sanitation and a better understanding of disease, it did not revolutionize treatment. Just because it was possible to see and grow bacteria did not mean that finding ways to kill them would be easy. Another pioneer, Paul Ehrlich, who worked in Koch's bacteriology lab, was searching for "magic bullets"— dyes, poisons, and heavy metals—that would stain specific germs and then, in a double whammy, attach to the germs and kill them.

But no one thought to look in the natural world for living organisms that would knock back pathogens. Why would they? The astonishing diversity of the microbial world is only now becoming appreciated.

Such was the mind-set of the scientific community when Alexander Fleming, a bow-tie-clad Scotsman working in St. Mary's Hospital in London, made a discovery that changed the world. Like many of his contemporaries, he was looking for ways to kill bacteria. In classically designed experiments, he placed a jellylike growth medium (agar and heated blood) into shallow, circular, transparent plates, called petri dishes, and then inoculated the medium with bacteria. Bacteria, which are too small to be seen with the naked eye, love to eat the nutrients added to agar. As they ate, they divided again and again. Eventually agglomerations of millions of bacteria formed a colony that could be seen by the naked eye. After putting the plates into a warm incubator overnight, Fleming was able to grow huge, clearly visible, gold-colored colonies of *Staphylococcus aureus* and other bacteria that he would try to kill with enzymes derived from white blood cells and from saliva.

In August 1928, Fleming went on vacation to France. When he returned in early September, he found several petri dishes that he had neglected to throw out. They had been inoculated with *Staph* and then they sat on his lab bench for the month he was away. As Fleming was tossing out the now useless plates, one of them caught his eye. It was flecked with a patch of blue-green fuzz, which he recognized as the common bread mold *Penicillium*. He saw that the luxuriant lawn of golden *Staph*, the filmy layer of billions of bacterial cells growing wall-to-wall on the plate, had disappeared near the mold. There was a kind of halo around the mold delineated by something in the medium that had prevented the *Staph* from growing.

Fleming's trained eye immediately recognized what had happened. The mold, which is a fungus that also likes agar, had produced a substance that had diffused into the agar and killed the *Staph*. That substance, the first-identified true antibiotic, dissolved bacterial cells, just as did lysozyme, an enzyme he had discovered in saliva in his experiments years earlier. It wiped them out, scorched earth–style. Fleming thought his "mold juice" contained an enzyme (like lysozyme), although it was later learned that the substance was not an enzyme but nevertheless disrupts the ability of bacteria to build their cell walls, causing them to burst.

The miraculous mold was identified as *Penicillium notatum*. Actually the antibacterial effects of *Penicillium* molds had been known since the seventeenth century but not to Fleming or any of his contemporary physicians. Ancient Egyptians, Chinese, and Central American Indians all used molds to treat infected wounds. But it was Fleming's training as a scientist that enabled him to move the fungus from a folk remedy into the scientific spotlight.

Over the next months, Fleming was able to grow the mold in liquid broth, pass the broth through a filter, and isolate a fluid that was rich in antibacterial activity. He called it penicillin. But there were many obstacles to obtaining enough of it. Not all strains of *P. notatum* made penicillin. Fleming was fortunate in that the one that fell onto his petri dish produced penicillin, but the yields remained tiny, unstable, short-lived, and slow acting. Unable to devise ways to make penicillin medically useful, Fleming gave up. After publishing his results, and trying some crude extracts on a few ill patients without any apparent effect, he concluded that his discovery had no practical importance.

But others had noticed. A few years later, a chemist in Germany who worked for the giant I. G. Farben chemical company that made aspirin and dyes used for textile coloring was looking for a dye that would inhibit the growth of bacteria. In 1932, Gerhard Domagk discovered a red dye (called *Prontosil*) that contained a wholly synthetic antibacterial agent, the first sulfonamide. A class of related sulfa drugs followed. These were the first agents that had any sustained and reproducible activity against bacteria and were not so overly toxic to people that they were injured by the side effects. Over the next few years, doctors began to use sulfa drugs to treat infections. But their spectrum of activity was limited. The drugs were good but not good enough.

With the outbreak of World War II, the need for antibacterial agents was urgent. Thousands of soldiers were destined to die from battle wounds, complicating pneumonias, and abdominal, urinary, and skin infections. In 1940 a team at the Sir William Dunn School of Pathology at Oxford University, led by Howard Florey and Ernst Chain, dusted off Fleming's penicillin and embarked on a journey to develop ways to make it in quantity. Because London was being bombed, they took their project to the Rockefeller Foundation in New York, where they were introduced to several pharmaceutical firms in the area. The companies did not welcome the scientists with open arms, for they knew that penicillin was at a very early experimental stage. Yields rarely exceeded 4 units per milliliter of culture broth—a drop in the proverbial bucket.

So the British scientists took their efforts to Peoria, Illinois, where the new Fermentation Division of the Northern Regional Research Laboratory was gearing up studies about using the metabolism of molds (fermentation) as a source of new microorganisms. Its staff was experienced and had a substantial collection of molds, but few of their strains made penicillin, and none was prolific. Thus the call went out to everyone they knew: send us samples of soil, moldy grain, fruits, and vegetables. A woman was hired to scour the markets, bakeries, and cheese stores of Peoria for samples bearing blue-green mold. She did the job so well they called her Moldy Mary. But in the end, a housewife brought in a moldy cantaloupe that changed the course of history. This particular mold produced 250 units of penicillin per milliliter of broth. One of its mutants churned out 50,000 units per milliliter. All strains of penicillin today are descendants from that 1943 mold.

The scientists ultimately developed methods for making this more potent form of penicillin in quantity. Later the pharmaceutical firm Charles Pfizer & Company used molasses as a way of growing the penicillium molds in bulk. By the time of the invasion of Normandy in June 1944, 100 billion units were being produced each month.

■ ■ ■

Penicillin ushered in a golden age in medicine. Here was a drug that could, at last, treat infections caused by deadly bacteria. Because it was so astoundingly effective, it was considered to be truly "miraculous." What could this wonder drug not do? Press reports heralded "a new era of medicine, the conquest of germs by interfering with their eating and digesting, [that] is sweeping through the military hospitals of America and England."

In 1943 streptomycin, the first effective agent against *M. tuberculosis*, was developed from soil bacteria. Then came others: tetracycline, erythromycin, chloramphenicol, and isoniazid, which together brought about the antibiotic era. At the same time, new forms of semisynthetic drugs were developed via chemical modification of natural substances, as well as the manufacture of purely synthetic or nonnatural compounds. Today, for convenience, we call all of these drugs antibiotics, although strictly speaking antibiotics are substances made by one living form to fight against another.

Those original antibiotics and their descendant drugs transformed the practice of medicine and the health of the world. Formerly lethal diseases like meningitis, heart valve infection, and childbirth fever could be cured. Chronic bone infections, abscesses, and scarlet fever could be prevented and cured. Tuberculosis could be arrested and cured. Sexually transmitted diseases like syphilis and gonorrhea could be cured. Even my case of paratyphoid could be cured without months of illness and a big risk of dying. Cure also was a great form of prevention, since a cured person would not spread the pathogen to others.

Surgery got safer. Antibiotics could be given pre-operatively to lower the risk of many surgical infections. If infection developed, antibiotics came to the rescue. Surgeons could attempt more sophisticated surgeries to correct a myriad of woes, such as removing brain tumors, correcting deformed limbs, repairing cleft palates. It is fair to say that without antibiotics there would be no open-heart surgery, organ transplantation, or in vitro fertilization.

Similarly, chemotherapies used to fight cancer often suppress the body's ability to fight infection and lead to bacterial infection. Without antibiotics, leukemias and many other cancers would not be treatable. It would be too dangerous to give the massive amounts of chemotherapy required without the safeguard of antibiotics.

In the 1950s the Chinese government decided to wipe out syphilis. Tens of millions of people were treated with a long-acting form of penicillin. This massive public-health campaign worked. The age-old scourge was virtually eliminated from China. Yaws, a related ancestral disease, was successfully eradicated from vast swaths of Africa after a series of similar campaigns.

How did and how do these drugs perform their miracles? Antibiotics work in three general ways. One, as exemplified by penicillin and its descendants, is by attacking the machinery used by bacteria to create their cell walls. With defective walls, bacterial cells die. Interestingly, they often commit suicide: lack of a cell wall triggers bacterial hara-kiri. We are not certain of the biological reason for their suicide, but nature selected for fungi like the *Penicillium* mold that make these antibiotics and are able to exploit that weakness.

The second mechanism is inhibiting of the way bacteria make

the proteins that perform all of the important functions of the bacterial cell. The proteins within a cell are vital for life. Cells need proteins to digest food, build their walls, enable reproduction, defend against invaders and competitors, and help the bacteria move around. Such antibiotics directly target the machinery that allows proteins to be made, crippling the bacteria, and having minimal effects on protein production by human cells.

A third is interfering specifically with the ability of bacteria to divide and reproduce, thereby inhibiting their doubling. With slower growth, they become less of a threat so that the host can mount an immune response to deal with them more easily.

If you think about it, antibiotics are natural substances made by living organisms—fungi and other bacteria—that want to throw a monkey wrench into the workings of their competitors. Their neighboring bacterial cells are little machines, all with multiple moving parts; over the eons, they have found many different ways to attack. And the bacteria have found so many ways to defend, which are the very basis of bacterial resistance to antibiotics. Since time immemorial, it has been an arms race. But for us humans, our development of antibiotics has been like getting the atomic bomb. It has fundamentally changed the playing field. Interestingly, both came on the scene at the same time. The scientific developments of the 1920s and 1930s led to their deployment in the 1940s. As with the atomic bomb, our hope was that it would be a panacea. The threat of the bomb would be so great that we would war no more. Similarly, the power of the antibiotics would once and for all defeat bacteria. Although there is some truth to both, neither has fulfilled that promise, nor could they ever. Both are just tools, and the fundamental causes of war between men and between men and bacteria remain.

■ ■ ■

As use of antibiotics became more and more widespread, a few side effects appeared, but most were mild—a few days of loose bowel movements, an allergic rash. In nearly all cases, these problems went away as soon as the drug was stopped. A handful of people had serious, sometimes fatal, allergies to penicillin. But the risk of dying from penicillin allergy was, and is, lower than the risk of being struck by lightning. It is a remarkably safe drug.

Other antibiotics did produce adverse effects. Some damaged the auditory nerve; others could not be used in children because it mottled their teeth. A very commonly used antibiotic in the 1950s, chloramphenicol, was found to cause a rare suppression of the bone marrow's ability to form blood cells, which was fatal about once in forty thousand courses of the drug. For very serious infections, such a low risk of dying from a drug allergy was infinitesimal compared to the risk of dying from the infection. But in some places, hundreds of thousands of healthy young children with mild sore throats were treated with chloramphenicol. For them, the risk clearly exceeded the benefit, and there were lots of alternative antibiotics. Doctors stopped using it almost entirely. Still, for years, I have said to my students that if I were marooned on a desert island and I could have only one antibiotic with me, I would choose chloramphenicol; it is that powerful.

The idea that other potent antibiotics also could have side effects beyond those immediately apparent wasn't part of the conventional wisdom; it was not even a consideration. If there were no allergies manifested in the days and weeks after receiving the drugs, they were considered safe.

Almost all the great advances in medicine from the second half of the twentieth century continuing through to today were catalyzed by the deployment of antibiotics. No harm could come from their use, or so it seemed. The fallout appeared only later.

6.

THE OVERUSE OF ANTIBIOTICS

How to capture the euphoria of those early days? It was 1945. World War II had ended. We had beaten the forces of great evil; a more just society had prevailed. Americans were awash with optimism. It was a time to have lots of babies, and I was one of them. In the five years after the war, Americans bought 20 million refrigerators, 21.4 million cars, and 5.5 million stoves. It was an era for all kinds of beginnings: Tupperware, the first fins on cars, sprawling suburbs, fast-food restaurants, television, and, of course, antibiotics—the wonder drugs.

Because they were so effective and so apparently free of obvious risk, doctors and patients alike began to ask: Can't we solve this problem with antibiotics? Like urinary tract infections? Can't we alleviate the discomfort of sinus or dental infections, or the disfigurement of serious acne? Can't we treat this condition with antibiotics? Like cystic fibrosis? The answer, very often, was yes.

Sometimes the benefits were dramatic, such as using antibiotics before certain surgeries as prophylaxis to prevent infections. Other times the benefits were small, but since the recognized cost, in terms of toxicity, was extremely low, even a marginal benefit seemed worth it. For example, for decades dentists routinely gave antibiotics to people with minor heart murmurs to prevent an exceedingly remote risk of heart valve infections.

I'm not questioning the efficacy of antibiotics on the small minority hospitalized with pneumonias, puerperal sepsis, meningitis, and other severe infectious diseases but rather their use on millions of healthy people with less serious infections and relatively minor complaints, such as runny noses and skin infections. Tens of millions of these people, year after year, are prescribed antibiotics in the United States alone.

The problem, as coming chapters will show, is particularly perilous for our children. They are vulnerable in ways we never foresaw.

The most obvious example of the extravagant use of antibiotics is for the common disorders known as upper respiratory tract infections. Parents of young children know the symptoms all too well: sore throats, runny noses, sniffles, earaches, sinus pain, and just plain misery. Sometimes their child has a fever, sometimes not. Most children get several upper respiratory infections every year until they are two or three years old. By age three, up to 80 percent of children have suffered at least one acute infection of the middle ear. More than 40 percent of kids experience at least six such ear infections by age seven.

Indeed, everyone—adults and children alike—develops upper respiratory infections with some regularity. We can't escape them. They are a product of our elaborate social networks. We are exposed constantly to plumes of microbes emitted from people coughing, sneezing, and just breathing all around us. These infections will be with us as long as we live in close proximity to one another, which is how most of us want to be: near our loved ones, our friends, our schoolmates. When scientists went off to spend winters in isolated colonies in Antarctica, upper respiratory tract infections would circulate for a month or two and then die out. As with the hunter-gatherers, everyone who was susceptible got sick, and then the infectious agent had nowhere to go. There were no new hosts to infect. Only when the next plane or next ship brought in new people with their communicable microbes would the cycle begin anew.

But guess what? Upper respiratory tract infections are mainly caused by viruses. More than 80 percent of them can be traced to microbes with exotic names like rhinovirus, astrovirus, metapneumovirus, and parainfluenza. (The term *computer virus* derives from these ultracontagious human viruses.) When we get infected with one of these pathogens, we say we have come down with a cold or the flu. But after a few days of feeling somewhere from "under the weather" to dreadful, nearly everyone gradually gets better. The condition is "self-limited." Even stubborn lingering coughs almost always fade naturally after a couple of weeks. But if you've been coughing for a week with no end in sight, you just might call your doctor and say, "Enough already. Give me an antibiotic." But in fact, antibiotic treatment has no effect on the outcome of these viral infections.

When it comes to infectious diseases, a central distinction is made between bacteria and viruses. Bacteria are cells. They eat, move, breathe, and reproduce. Give bacteria their preferred nutrients and a nice home—which could be a warm nook, glacier, or volcano—and they will multiply.

In contrast, viruses are much smaller and simpler. They require a host. They can only live within a cell, be it from a human or other animal, plant, or bacterium. They hijack host cell machinery for their purposes, which includes reproduction. They cannot reproduce on their own. Sometimes they stay dormant in host cells for decades, and at other times they kill their host cells—or do both.

Since viruses don't have cell walls the way bacteria do, antibiotics like penicillin don't work on them. Since their protein synthesis is dependent on their host's protein synthesis, you would have to inhibit the latter to suppress the virus. When viruses parasitize human cells—as they do with the common cold, herpes, flu, and many other infections—we cannot suppress host protein synthesis because the host is we humans. We would be poisoning our own bodies. A few drugs do interfere with the ability of some viruses to enter and exit cells or reproduce, such as acyclovir, used to treat herpes viruses, and drugs that interfere with the life cycle of HIV. Although viruses can be suppressed, there are few cures. Antibiotics, in contrast, can cure almost all bacterial infections.

But fewer than 20 percent of upper respiratory tract infections are caused by bacteria. Here, the situation gets more complex. Bacteria that occupy your throat and nose can be permanent residents, or they can be transients, or they are in-between occupants, akin to long-term tenants. Some of the most important include *Streptococcus pneumoniae* or the pneumococcus, the number-one pathogen in the upper respiratory tract and in the lungs that causes ear infections and pneumonia, respectively; *Streptococcus pyogenes* or Group A strep, which causes "strep throat"; *Staphylococcus aureus,* the bacterium found in most serious *Staph* infections; and *Haemophilus influenzae,* which used to cause ear infections with regularity and occasionally childhood meningitis before we had a vaccine.

These four bacterial species are frequently identified in upper respiratory tract infections. But not so fast. Sometimes they cause the infection, but most of the time they do not. This apparent contradiction can be explained by the fact that these microbes, whose names sound ominous, may have actually arrived in you or your child quite some time ago. You have not been infected, but instead colonized, a generally harmless event. This is an extremely important distinction that often gets overlooked.

Colonization means that these bacteria are merely living in and on you. They are not doing you any harm. While colonization is a prerequisite for most disease, by itself it is not sufficient. Most colonized people are perfectly healthy. For instance, *Staph aureus* can colonize your nose for life, and you'll never know it. For most people, it is a part of our microbiome, as discussed previously. The bottom line: our noses and throats are hosts to vast communities of bacteria, including the friendly ones and those that are potentially pathogenic.

Moreover, certain of these bacteria have been shown to keep you healthy by inhibiting potential pathogens and modulating your immune system. One of the most interesting examples of this is the "viridans" streptococci, a group of bacteria that live peacefully in everyone's mouth. They were initially labeled as pathogens because they were a leading cause of heart valve infections. But it gradually became clear that they are normal residents of the mouth, only occasionally entering the bloodstream and landing on a previously damaged valve. We now know that if we mix harmless viridans with pathogenic Group A strep, viridans always win out. They knock back the strep. So a bacterium that is an occasional pathogen turns out to be a significant protector of our health. This dichotomy is an important model for how to think of many other bacteria that commonly live in us.

But getting back to the main point, when do potential respiratory pathogens cause trouble in children? When should they be treated with antibiotics? One answer is another question: How healthy are your children? If they have another infection, like the "stomach flu," or stresses on their immune system, like an allergy that blocks their Eustachian tubes, they are more vulnerable to developing more serious ear or chest infections. In rare instances, these can lead to serious complications such as pneumonia or mastoiditis, an infection of the air spaces adjacent to the ear canals.

Infections can also linger in apparently healthy children. If a thousand kids in your town have been exposed to the same respiratory virus or bacteria—not an unlikely occurrence during winter—we see a range in outcomes. Some kids have no symptoms whatsoever; they are simply carriers. Others have symptoms for one day, others still for two or three days. After four or five days, the numbers drop, but a few will always have prolonged infections. The distribution falls along a familiar bell-shaped curve: some not sick, most sick in the normal pattern, a few very sick.

A doctor can recognize a severe infection but cannot easily predict who will have trouble recovering. Thus while the numbers of the very sick are low, maybe 5–10 percent of those affected, 60 or 80 percent of kids who currently are taken to the doctor by their parents complaining of bad sore throats or ear pain walk out with an antibiotic. And most of the time the doctors have no idea if the illness is caused by bacteria or a virus.

Doctors have one very good reason to reflexively prescribe antibiotics for many upper respiratory tract infections: fear of rheumatic fever. It is a very serious inflammatory disease, resembling rheumatism, which typically occurs two to three weeks after an untreated strep infection (*Streptococcus* pharyngitis). Antibodies raised against the strep infection "cross-react" with and attack the child's heart muscle, joints, skin, and brain—a tragic case of mistaken identity.

Before antibiotics, about one child in three hundred with a strep infection developed rheumatic fever or, if the strep strains were very "hot," one in thirty. Nowadays doctors prescribe an antibiotic for strep throat not to shorten the duration of the infection, because it doesn't much, but to ward off rheumatic fever. Most people and even some doctors don't understand that the antibiotic is being prescribed for prevention and not treatment.

But here's the problem. Kids get colonized with Group A strep in their throats all the time, especially in winter. The condition may persist for a couple of months, as they are just healthy carriers. But imagine your child acquires an ordinary cold virus during this same time and gets a sore throat. You take him or her to the doctor, who does a throat culture, and voilà: Group A strep shows up. The doctor, being reasonable, prescribes antibiotics to prevent rheumatic fever when the infection is actually caused by a virus.

Even if the strep throat is caused by bacteria, the illness is usually brief, and nearly all kids get better in the next day or two. But if your child receives antibiotics and gets better, you will inevitably think the drug made them better. It's a classic example of the adage that correlation is not causation. The fact that your child improves after several doses of amoxicillin seems clearly correlated, but that does not prove the drug led to the improvement.

How, then, can physicians tell the difference between a mild, self-limiting bacterial or viral infection and a more serious one? Or how can they differentiate colonization from infection? This is a critically important question because the answer, which unfortunately is not clear at this time, holds the power to help curtail the overuse of antibiotics. An astute clinician knows that in most, but not all, cases, children who are at risk for developing serious complications show certain warning signs. They have higher fevers, their symptoms have been going on longer, their white blood counts are more abnormal, and they look worse. But many cases fall into a gray area.

This gray area is important. Until doctors can readily distinguish viral from bacterial throat infections, they will always follow the safer course. And they are pressed for time. They may have to see five sick children every hour of their working day and complete all of the paperwork. The conjunction of lack of practical, rapid, inexpensive, and accurate diagnostics and the incessant time crunch effectively conspire on the side of overtreatment. New diagnostics that can improve the situation are coming along, but they are hardly ever deployed; in the current climate, no one wants to pay.

And then there is always the fear of the lawyer looking over the doctor's shoulder. What if the physician doesn't treat a child and there is a catastrophic outcome? What if an attorney asks: "Why didn't you give this child an antibiotic for his ear infection that turned into meningitis and left him paralyzed?"

The complicated dynamic just described has been playing out on an unprecedented scale, involving all of our children all over the world for generations. The cycle is repeating, perhaps amplifying. When millions of children are treated for bacterial infections they never had, it's not hard to imagine that trouble could follow.

The magnitude of antibiotic use is enormous, and it has crept up year after year. In 1945 an article in the prestigious *Journal of Clinical Investigation* reported on the great efficacy of penicillin in treating sixty-four patients with pneumonia. Such treatment, on that scale, produced almost miraculous results. But by 2010, health-care providers prescribed 258 million courses of antibiotics to people in the United States. This more than a millionfold difference in scale amounts to about 833 prescriptions for every thousand people across the country. We don't know if every course was taken, but probably most were. Family practitioners prescribed about a quarter of the antibiotics, followed by pediatricians and internists. Dentists prescribed 10 percent, about 25 million courses a year.

The highest prescription rate was for children under the age of two: 1,365 courses per 1,000 babies. This means that the average American child received nearly 3 courses of antibiotics in his or her first two years of life. They go on to receive, on average, another 8 courses in the next eight years. Extrapolating from the current Centers for Disease Control and Prevention statistics, the data suggest that on average our children receive about 17 courses of antibiotics before they are twenty years old. This is a big number, but it is in line with prior studies in the United States and other developed countries.

Young adults in their twenties and thirties receive, on average, another thirteen courses of antibiotics. This means that our young people are taking thirty courses of these potent drugs before the age of forty. This is the average. Some people take more, some less. But the implications loom large. Many of the young women will be mothers of the next generation who will be providing initial microbiomes to their children. This is a process we will explore shortly. How will all of those courses of antibiotics affect the handover?

■ ■ ■

The first recognized problem from the overuse of antibiotics was resistance. Very simply, the more often we put antibiotics into our bodies and our children's bodies, the more likely we select for bacteria that are resistant to their action. Many people don't quite understand resistance. They think that "they have become resistant to antibiotics," when in fact it is the bacteria they encounter or they carry that have become resistant.

Here is one way it works. A child receives an antibiotic, say amoxicillin, to treat an infection. A derivative of penicillin, it is the most commonly prescribed antibiotic in young children in most countries. When amoxicillin is swallowed (usually in the form of a bubblegum-pink liquid), it is absorbed into the gut and enters the bloodstream. From there, it travels to all organs and tissues, including stomach, lungs, mouth, throat, skin, ears, and in girls the vagina, encountering and destroying bacteria wherever they hang out. So-called broad-spectrum antibiotics like amoxicillin are especially adept killers.

But here's the crux of this problem: there are always innocent bystanders, lots and lots of them. All mixed populations of bacteria include both susceptible and resistant bacteria. The antibiotic elimi-nates susceptible microbes all over the body along with the pathogen that usually is present in one place. It is like carpet bombing when a laserlike strike is needed.

And now we are in trouble. When susceptible species are dimin-ished or killed, populations of resistant bacteria expand. With fewer competitors around, resistant bacteria flourish. They're the lucky ones that go on to have lots of baby bacteria. The resistant ones may be either the targeted pathogen or the many, many bystanders.

Antibiotic resistance spreads within bacterial communities in two general ways. First, it occurs through the growth of organisms that have already acquired resistance—what we call vertical transmission. It's like the transfer of genes from grandparents to parents to children and so on down through the generations. When antibiotics are in the environment, bacteria that are resistant behave in a similar manner. They keep dividing and multiplying, passing on their genes, unlike susceptible bacteria, which are inhibited or killed.

Resistance genes can also spread via sex—what we call horizon-tal transmission. Some bacteria are reclusive, but many bacterial spe-cies are promiscuous, having sex all the time. But it's not exactly as you might picture it—two bugs lying on a couch going at it. Instead, they may gain or swap genes like baseball cards, many of which con-fer resistance to antibiotics. The genes encode enzymes that modify and inactivate antibiotics or encode proteins that pump the antibiotic out from the bacterial cell. When resistance genes are present and antibiotics are around, there is natural selection for the strains that have resistance genes to propagate. Thus, the surviving bacteria may be said to have adapted to the antibiotics designed to kill them, rendering the drugs less effective or completely useless. As long as antibiotics are around, resistance is favored in the populations of microbes.

The dynamics of how resistant bacteria emerge are instructive. For example, a small dose of amoxicillin is enough to kill nearly all the pneumococcus encountered but not quite all. In a population of a million pneumococci, there might be one with a small genetic variation—an outlier—that arose by chance and that is resistant to amoxicillin. After 999,999 of the others are killed, the variant multiplies, basking in the empty niche that the amoxicillin created. It becomes dominant. Occasionally, one of these resistant bacteria is passed to another child through a cough or sneeze. Now let's imagine that the second child also gets a high dose of amoxicillin. Again all of the susceptible pneumococci die. And now, from among those more-resistant bacteria, a variant that is even more resistant survives and thrives, armed with its usual bacterial weaponry. And so on.

Resistance ramps up, little by little, or it can happen quickly. Sometimes, a resistant strain picks up new genes from another bacterium by having sex with them and, in a single bound, acquires high-grade resistance to a whole class of antibiotic agents. Many times that gene was acquired from a bystander organism that had been selected and enhanced by prior courses of antibiotics.

As long as amoxicillin is given to our children who have pneumococci in their noses and throats, whether harmless or not, antibiotic resistance is inevitable. It does not happen in every child or with every course of the drug. Sometimes variants don't arise, or they do but they are less fit and are not transmitted to other kids. It is a kind of casino; in any individual child, or in any community, chance plays a big role. The resistant ones may fail and may be eliminated; this probably happens most of the time. But some may persist for years. In a later chapter, I discuss our studies showing this. But globally, in just this manner, resistance to penicillins has slowly and inexorably mushroomed in recent decades. It is but one example, since resistances to macrolides (like erythromycin, clarithromycin, and azithromycin), tetracyclines (like doxycycline), fluoroquinolones (like ciprofloxacin), and nitroimidazoles (like metronidazole) are all rising.

One issue is that parents are not aware of, or don't much care about, antibiotic resistance as it develops in the broader community. Going back to the example of ear infections, the conversation in the doctor's office might go something like this.

DOCTOR: The reason that your daughter is so fussy is that she has an ear infection.

MOTHER: I figured as much, since she's had them before. Can we give her an antibiotic?

DOCTOR: Well, in more than eighty percent of cases, the infection is due to a virus, so antibiotics won't work.

MOTHER: What about the other twenty percent?

DOCTOR: Well, we are overusing antibiotics. The more we use them, the more resistance there is, and resistance is spreading across the community.

Mother does a quick calculation. The community means other children. But her child could be among the 20 percent: "Antibiotics won't hurt, and I want to do whatever we can for her."

Doctor does another quick calculation. It's true: antibiotics might not help her, but they won't hurt: "Okay, I'll prescribe a course of amoxicillin for ten days."

■ ■ ■

A second crisis is looming, accentuated by our overuse of antibiotics and resistance to the drugs: the failure of pharmaceutical firms to develop new antibiotics to keep up with the resistance. Some infections today are untreatable with current antibiotics, and more are likely to evolve.

Antibiotics vary from narrow spectrum, those affecting only a few types of bacteria, to broad spectrum, those that kill a wide variety of microbes. Most pharmaceutical companies favor broad-spectrum drugs because the broader their use, the greater their sales. Physicians also like them and for good reason—because it can be difficult to determine whether an infection is caused by *Strep*, *Staph*, or *E. coli*, and broad-spectrum agents cover the waterfront. But there is a significant downside: the broader the spectrum, the greater the selection for resistance.

It is clear that the more we use antibiotics, the quicker resistance will develop and the useful life span of each antibiotic will be reduced. In the early days of antibiotic discovery, scientists generally stayed ahead of this curve by regularly developing new drugs. But now the pipeline is drying up. The "easy" antibiotics have already been discovered. Like putting different colors of frosting on the same cupcakes, most drug companies have tweaked their existing antibiotic recipes without coming up with new ingredients.

It is not profitable for companies to go to the trouble and enormous expense of developing new antibiotics, especially if they don't have broad application. Pharmaceutical companies want to develop drugs that millions of people will take for years, such as medications to treat high cholesterol, diabetes, and high blood pressure. That is where the profits are. A drug that only a few thousand people need each year and that will be used for only a few weeks can't be developed under our current economic models.

A few years ago, when I served as an officer of the Infectious Diseases Society of America (IDSA), one of my jobs was to participate in efforts to convince the U.S. Congress to pass laws that could help us open up the stalled antibiotic pipeline. We at the IDSA were (and still are) very concerned about the lack of new drug development and knew that the process takes years. We cannot wait for a highly communicable bacterium that is resistant to all of our antibiotics to hit before taking action. I traveled to Washington repeatedly over several years to work with other IDSA team members, other organizations with the same goals, and the family members of people who had died or became very ill as a result of resistant organisms. We testified in Congress whenever we had the opportunity, in briefings or in formal committee meetings.

The stories of young, healthy people struck down by terrible, unrelenting infections were terribly sad and frightening. One day, Brandon Noble, a professional football player who played for the Washington Redskins, came to testify. He had reached the top of his field and was familiar to everyone in the room. But like many professional athletes, he had sustained a series of injuries, which in his case damaged his knee. He went to the hospital to have the torn ligaments repaired, a relatively routine operation. Thousands are done without mishap each year. But his knee became infected with antibiotic-resistant *Staph*, called MRSA (MRSA stands for methicillin-resistant *Staphylococcus aureus*). His knee had to be drained and cleaned out multiple times; despite these necessary treatments, the moving parts of his knee were permanently mangled. When his infection was finally cured, he could no longer walk normally, and his career was over. Limping up to the microphone, the extent of his loss was immediately evident. He later said: "The worst and most unexpected thing that I have come up against in my football career has been a tiny little thing that I can not see."

The next witness was a mother from a small town in Pennsylvania who spoke about her son, Ricky Lanetti, a college senior who also was a football player. He was getting ready for the play-offs for the NCAA Division III championships when he noticed a sore area on his rear. It was a small abscess, pretty routine, only a little bigger than a pimple. No one, least of all him, was very concerned; he was preparing for the big game.

Within days, this young man died of a severe MRSA infection that spread from the abscess throughout his body. His immune system could not contain it, and no amount of antibiotic could save him. His mother's grief resounded in the now-silent room. She showed me a beautiful picture of her standing with her son in his football uniform towering over her; now he was gone.

When Congress members consider any given issue, they sometimes invite a few interested parties to a panel convened by one of the subcommittees in the Senate or the House. The meetings are held in large rooms, impressive in their classic structure and furnishings and as symbols of the power of our democracy. The rooms are filled with people in pecking order—the congressional panel, sitting in the front on a dais, the tables before them where the speakers will testify, and then the seats in the back, where the waiting speakers sit along with congressional aides and others who want to hear the action.

A hearing often has three or four panels of speakers, organized by the staff in terms of the particular items to be discussed. Congress members and celebrities come first, then their friends, then the organizations that have an interest. I testified multiple times on this issue, and the IDSA, despite being the professional organization most concerned and most knowledgeable on the subject, was always in the last panel. By then, after hours of numbing testimony, self-congratulatory speeches by both testifiers and congressmen and -women, and breaks, the room was usually almost empty. Most of the Congress members had departed, but still the chair had to be there to preside, to conclude the nation's business.

This exact scenario played out yet again. Finally I was seated at the table for testimony. I had readied my speech about why we needed to strengthen the pipeline and our suggestions about how to do it. The only congressman still there was the subcommittee chair, an older man with a distinctly southern accent. Before I could start, he said that he was glad to hear testimony about this. He continued: "A few weeks ago, I was with my friend. We were golfing. He was telling me how much his knee was hurting and how he was scheduled for surgery for a knee replacement. The next time I saw him was at his funeral. From the surgery, his knee became infected with the MRSA, and it killed him, just like that. There was no good way to treat him. So I know what you are talking about."

There were only a handful in the room to listen, but the congressman had captured the essence of why we need to do something. His committee reported favorably about the legislation, which eventually became part of a federal act to create incentives for companies to develop new antibiotics. However, the paradox remains that we use far too many antibiotics, but we don't have enough of the right ones to treat an emerging burden of these antibiotic-resistant infections. Actually, the problems are related; the first drives the likelihood of the second.

But antibiotic resistance is not just a problem of people overusing antibiotics. The problem also stems from how we treat animals down on the farm.

7.

THE MODERN FARMER

Imagine cattle grazing peacefully in a pasture, chewing their cud, moving from place to place to eat fresh green grass. You might envision a Norman Rockwell scene of our agrarian past: carefully tended barns, beautiful picket fences, contented cows, and, breaking the silence, the buzzing of an occasional fly and flapping of tails.

Here's another picture: cows lined up in small metal pens, row after row of them, with their heads braced into corn-filled troughs. A dense, pungent odor of cow manure wafts from miles away. Cows are released into vast feedlots where they mill around on bare ground, eating all the time, surrounded by their poop.

Most antibiotics produced in the United States go not to humans but to these massive feedlots and their equivalents for swine, chickens, and turkeys. They are modern, integrated, industrial operations for fattening up millions of animals for slaughter, billions in the case of chickens. Agricultural science operates to maximize meat production, with a particular focus on optimizing feed efficiency—the conversion of calories in animal feed into meat. Feeding antibiotics to farm animals plays a central role in that process, fattening them up. But it has also led to antibiotic resistance in the microbes inhabiting livestock and to antibiotic residues in our food and water. And it provides an important, albeit distasteful, analogy about what we might be doing to our children.

We now know that antibiotic resistance develops in humans when a drug kills susceptible microbes while sparing the occasional microbes that through genetic variation have acquired resistance. The resistant species proliferate, making further antibiotic treatments less effective. The same thing happens on the farm, but here I will elaborate on the process in more detail.

Bacteria, fungi, and algae have been fighting one another for hundreds of millions of years to gain the upper hand in an endless game of chemical warfare. In their struggle to survive, they produced natural antibiotics for self-defense. At the same time, they evolved genes that would counteract their own antibiotics and those of their enemies. Thus, two classes of sophisticated genes emerged in microorganisms: those involved in making antibiotics and those that provide resistance to antibiotics.

In 2011 scientists analyzing thirty-thousand-year-old bacteria from the Yukon permafrost found that these bacteria were able to resist antibiotics, both those occurring naturally as in bread mold and the semisynthetics, which have similar core structures. The finding provides direct evidence that ancient genes for antibiotic resistance were widespread and long preceded our human use of antibiotics for treating illness. One implication of the ancient arms race is that we didn't cause resistance. To say that it is not our fault is only partly true; although resistance is ancient, we have made it a lot worse. We don't even know how many orders of magnitude we have multiplied it, within our human space, but it is surely considerable. Even ocean life, offshore but living in our wastes, shows evidence of the spread of resistance from our activities. It is a fingerprint that we are leaving everywhere.

Another implication of the ancient nature of resistance is that there will be no easy solution to the problem. We will never make resistance go away, because Darwin was correct in his theories. There always will be strong selection for resistance when populations encounter stress, in this case, microbes under antibiotic pressure. A corollary is that we will never invent a superantibiotic that cures everything. Microbes are too diverse, and Nature will always come up with new ammunition.

■ ■ ■

Our bucolic barnyards have been replaced by those feedlots and henhouses containing tens of thousands of animals. A single barn from a large hog operation can hold two thousand or more pigs. A single henhouse can hold twenty thousand or more chickens. By packing the animals into small, unsanitary spaces, farmers set up the perfect conditions for bacteria to proliferate and spread.

But the main reason farmers give their animals antibiotics isn't so they can crowd them together with fewer diseases. In fact, they don't give their animals full therapeutic doses, the levels that would be used to treat infections. On most factory farms, animals are given food or water treated with a low, subtherapeutic dose of an antibiotic. Farmers do this to enhance feed efficiency. This effect of subtherapeutic antibiotics is called growth promotion.

The practice dates back to the mid-1940s when pharmaceutical manufacturers discovered that animals fed antibiotics gain more weight in terms of muscle mass more quickly than do animals fed a drug-free diet. In reviewing the older literature, I found a particularly interesting study from 1963. Amazingly (at least to me), the nature of the interplay between gut microbes and antibiotics was described way back then. These scientists asked themselves whether the observed growth promotion effects in animals were due to antibiotics themselves (acting on tissues) or to their effects on the microbiome (which they called the normal flora). So they raised two groups of chickens: one in the usual circumstances, what we call conventionally, and the other under germ-free conditions. Germ-free animals have been raised so that there are no microbes at all living in or on their bodies. In each group, half the animals received antibiotics in their feed while the other half, serving as controls, did not.

As expected, conventionally raised chickens exposed to low-dose antibiotics grew bigger than the untreated controls. But the two groups of germ-free chickens turned up a surprise: those given antibiotics were no bigger than the ones who weren't. This suggested that a chicken's microbes are essential for the growth-promoting effect to occur; antibiotics alone were ineffective. These findings have been available for more than fifty years, but they were ignored and then forgotten.

The upshot is that farmers quickly realized that their animals could gain an extra 5, 10, or 15 percent of their normal body weight at a relatively low cost. And the corollary to this was that they gained more weight per unit of food that they consumed. This was called improved feed efficiency. The pharmaceutical companies found that they could make big profits by selling antibiotics to farmers by the ton rather than by the milligram to physicians.

Today, an estimated 70–80 percent of all antibiotics sold in the United States are used for the single purpose of fattening up farm animals: hundreds of millions of cattle, chickens, turkeys, pigs, sheep, geese, ducks, and goats. In 2011, animal producers bought nearly 30 million pounds of antibiotics, the largest amount yet recorded, for their livestock. We don't know the exact number because these amounts are closely guarded secrets. Both the agricultural and pharmaceutical industries are defensive about their practices. According to the former chairman of the Food and Drug Administration David Kessler, until 2008 Congress didn't require drug companies to tell the agency the quantities of antibiotics they sold for use in agriculture. Nor do the firms provide information on how drugs are administered or to which animals and why. Industry lobbyists have successfully blocked most attempts to curtail antibiotics in animal feed. And because of this ongoing battle, there has been little research on the pros and cons of growth promotion. With the exception of a few industry-oriented scientists, few people have been paying much attention.

Meanwhile, ecologists and medical doctors bemoan the practice of growth promotion, noting that farmers give animals the same drugs that people get from their doctors and surgeons. In 2013, Consumers Union tested hog carcasses and found that 13 of 14 *Staphylococcus* samples isolated from pork were resistant to at least one antibiotic. So were 6 of 8 *Salmonella* samples and 121 of 132 *Yersinia* samples. One sample was found to have MRSA, the fearsome drug-resistant and sometimes fatal form of staph that we discussed. Why would we squander our precious antibiotics, including the ones that save lives when nothing else helps, to make meat a few cents a pound less expensive?

In 2011 more than half of samples of ground turkey, pork chops, and ground beef collected from supermarkets for testing by the federal government contained bacteria resistant to antibiotics, what we sometimes call superbugs. Actually there is no such thing as a superbug (a term invented by reporters), but if one of these highly resistant organisms were to attack and infect your knee or heart valve and there were no effective antibiotics, you would readily think it had superpowers.

Resistance is not the only problem. The National Antimicrobial Resistance Monitoring System (a joint program of the Food and Drug Administration, the Department of Agriculture, and the Centers for Disease Control and Prevention) found that some 87 percent of the meat collected from supermarkets contained either normal or antibiotic-resistant forms of enterococcus bacteria, which are an indication of fecal contamination. Two such species of the bacteria, *Enterococcus faecalis* and *Enterococcus faecium*, are leading causes of infections in the intensive care units in U.S. hospitals. One possibility is that some of those patients acquired resistant versions of them from their food.

Sweden banned the use of antibiotics for growth promotion in 1986. The European Union forbade the practice in 1999. Since then, the use of all antibiotics in animal feed for growth promotion has been banned in all of Europe.

American food producers and pharmaceutical companies argue that there is no solid proof that antibiotic-resistant microbes from animals are infecting humans. But in fact we have evidence going back more than thirty years showing the same organism—with the same patterns of antibiotic resistance—turning up in sick people and in animals fed growth-promoting antibiotics. For example, more than two thousand different strains of *Salmonella* have been typed and have names, so we know who they are. A long series of *Salmonella* outbreaks in people have been traced back to factory farms. The microbes isolated from the animals, the food, and the infected people were shown to have identical molecular profiles as well as patterns of antibiotic resistance.

The stonewalling defies reason and represents the kind of hands-off libertarianism that is eroding our public health. Bacteria don't respect political dogma any more than they recognize political boundaries and jurisdictions. In March 2013, a Danish study provided yet another smoking gun. Using whole-genome sequencing of bacteria, researchers demonstrated that the MRSA infections of two Danish farmers were caused by the same organism that infected their animals, which could not happen by chance, providing evidence that they became infected by the strain from contact with their animals.

■ ▩ ■

The problem is not limited to the resistant bacteria that come to us in the food we buy from industrial farms. The antibiotics themselves arrive in our food, particularly in meats, milk, cheeses, and eggs. The Food and Drug Administration requires farmers to establish a wash-out period between giving the last dose of an antibiotic and when the animal is slaughtered. But inspections are infrequent and enforcement is minimal.

Foods on our supermarket shelves are allowed maximum residue limits for antibiotics, establishing upper boundaries of what is permitted. For example, milk can legally have up to 100 micrograms of tetracycline per kilogram. This means that a child who drinks two cups of milk a day will ingest about 50 micrograms of tetracycline every day. That is not a lot, but consider the fact that many children drink milk every day, year after year. And that level is just for tetracycline. All other antibiotics have their own allowable limits. A 1990 report indicated that 30–80 percent of milk samples had detectable antibiotics, especially sulfa drugs and tetracycline.

Surveys in the 1980s and 1990s showed that legal limits were exceeded 9 percent of the time in meats, milk, and eggs. Thus, you are likely to be ingesting antibiotics whenever you eat nonorganic meats, milk, or eggs. Most people who say they have not had an antibiotic in years are mistaken. Millions of us are exposed every day and not only from foods. Antibiotics are found in our water, especially around farm runoff and treated human sewage. Current water purification treatments are excellent for reducing harmful bacteria and viruses but they do not fully remove antibiotics. A 2009 study of several cities in Michigan and Ohio found antibiotic-resistant bacteria in all source waters, drinking water from treatment plants, and tap water. The amounts were small, with the greatest levels in tap water. The point is that it all adds up.

Densely farmed commercial fish, such as salmon, tilapia, and catfish as well as shellfish like shrimp and lobster, are given relatively high doses of antibiotics, not so much to promote growth as to combat diseases associated with the crowded conditions in which they are raised. As with livestock, the FDA requires a washout period, but fish raised domestically are rarely inspected. Fish and shellfish raised in Asia are more tainted. Violations abound.

The antibiotic oxytetracycline—closely related to a form of tetracycline widely used in people—and streptomycin are even used on organic apples and pears to combat fire blight, a bacterial disease of fruit trees. The use of such drugs does not have to be divulged. You probably never imagined that produce labeled organic could contain antibiotics. Drug-resistant bacteria also end up in fertilizer and soil, further contributing to the reservoir of resistance in our ecosystem.

Modern agriculture, with the intensive production of everything from livestock to fruit, is bringing antibiotic-resistant bacteria and the antibiotics themselves directly to humans. Later, we will discuss the possible consequences. But in terms of my work, the most important aspect is growth promotion. If receiving antibiotics at a young age fattens up our farm animals, changing their development, then might that be analogous to what happens when we give our children antibiotics? Are we inadvertently fattening them up, when our intention is to treat their illnesses?

8.

MOTHER AND CHILD

In the 1950s two new drugs became available to treat common problems in pregnancy. One was thalidomide and the other was diethylstilbestrol, otherwise known as DES. These were drugs that were considered safe for pregnant women and had actual or purported benefits. Each provides a strong cautionary tale of the dangers of treating millions of healthy pregnant women with potent drugs.

First is the now infamous story of thalidomide. Discovered in East Germany in the mid-1950s and released in 1957 as a treatment for insomnia and anxiety, the drug was soon found effective in alleviating morning sickness. Women were thrilled. Essentially no one questioned its use because most scientists and doctors believed that drugs did not pass through the placenta; so if the mother was okay, the baby would be too.

Sadly, most of us know what happened. Between 1957 and 1961, thousands of women were prescribed thalidomide. In 1960 it could be purchased over the counter in Germany without a prescription. Even today, we don't know how many women were exposed. What we do know is that at least ten thousand to twenty thousand babies were born with serious birth defects, mostly involving limb development, with shortened or missing arms and legs, as well as anomalies of the pelvis, eyes, and ears. Many defects were lethal. Once it became clear what was going on, thalidomide was immediately banned.

Luckily, Francis Kelsey, the commissioner of the U.S. Food and Drug Administration, withheld approval of thalidomide until it could be shown to be safe. Thus women in America were largely spared the tragedy of those birth defects in their babies, unless they happened to get thalidomide in another country. The toxicity of thalidomide was obvious at birth, so after multiple cases appeared, it was not so difficult to understand what had happened. Still, it took a few years, amid discussions and questions of whether the birth defects were due to nuclear bomb testing and other causes, to ban the drug. During those years, the toll of misery mounted relentlessly.

A second cautionary tale concerns a form of estrogen, DES, that was developed at Oxford University in 1938 under a grant from the Medical Research Council in England. It was never patented because of a policy against making profits on drugs discovered using public funds. As a result, DES was available to any company that wanted it, and many did. In 1941 the Food and Drug Administration permitted its use for treating a variety of menopausal symptoms, for stopping lactation after birth, and for reducing breast engorgement. DES had no obvious or important side effects, and in the 1940s, amid a great wave of enthusiasm, doctors began using it to treat pregnant women for a wide variety of problems, including the prevention of recurrent miscarriage and alleviating morning sickness.

DES came along at a time when the public believed in the power of medical science and trusted doctors' authority. Medical journal advertisements showed beautiful babies with great complexions, alert and smiling, implying that their robust health was due to DES. It was difficult for many physicians to resist the tide, since so many of their colleagues were using it, and large, reputable companies were promoting it. Probably more than 3 million pregnant women received DES, mostly in the United States but in other developed countries as well. Unfortunately, there was no real science backing up this faith in the drug. Its popularity was an exercise in pure marketing.

In 1953 a carefully conducted clinical trial was published in the *American Journal of Obstetrics and Gynecology* showing that DES did not improve pregnancy outcomes in the least. Gradually medical textbooks all came to say that it was not effective. Still, DES continued to be used in pregnancy for years afterward. There was a disconnect between what the medical literature was advising and what doctors were doing. Inertia, custom, and peer pressure prevailed. Even though it was ineffective, no one thought that it wasn't safe.

The first indication of trouble came in 1971, when doctors in Boston published a study concerning a very rare cancer called clear-cell adenocarcinoma of the vagina. Most vaginal cancers occur in older women, but these all occurred in adolescents or young adults. An investigation revealed that the mothers of seven of the eight patients in the study took DES in pregnancy. These girls and young women had been exposed to DES when they were in their mother's womb, but the consequences did not manifest until fourteen to twenty-two years later. More cases followed. We now know that having been exposed to DES in utero increased the risk of these cancers fortyfold.

While these are rare tumors, it turns out they were the tip of the iceberg. A 2011 study headed up by Dr. Robert Hoover at the National Cancer Institute compared the cumulative risks in women exposed to DES in utero with those not exposed and found a doubling of their infertility rate (33.3 percent vs. 15.5 percent). The DES babies had less of a chance to have their own babies. Exposure also had significant effects on their loss of second-trimester pregnancies (16.4 vs. 1.7 percent), higher rates of preterm deliveries with all of their attendant problems, and more cases of early breast cancer.

Sons of women who took DES also had heightened disease risks, including problems in their male genital tracts, such as cysts and failure of the testes to descend properly from the abdomen. A wisp of evidence suggests there may be similar effects in the grandchildren of women who took DES.

These dreadful health problems were not detected earlier because, unlike with thalidomide, the effects were delayed by decades. Also there are multiple reasons a woman can be infertile. Someone had to have a hypothesis and look carefully to see that cumulative risks for such problems were higher in DES babies. And now we know.

One takeaway from these stories shouts out to me. It's a lesson many of us learned earlier from our parents: just because everyone else is doing something doesn't mean that it is safe. Back then it was normal for pregnant women to receive DES and thalidomide. Today it is normal for women to have Cesarian sections and to take antibiotics during pregnancy. These practices are occurring on an unprecedented scale.

■ ■ ■

Throughout the animal kingdom, mothers transfer microbes to their young while giving birth. Different species of tadpoles acquire specific skin bacteria from mother frogs even though they all live in the same pond with the same bacterial background. Emerging chicken eggs get inoculated with microbes from a bacteria-filled pouch near the mother hen's rectum. And for millennia, mammalian babies have acquired founding populations of microbes by passing through their mother's vagina. This microbial handoff is also a critical aspect of infant health in humans. Today it is in peril.

In the past one hundred and fifty years, birth practices have changed dramatically. To be sure, the act of childbirth is safer than ever. Hospitals are equipped to handle the kinds of emergencies that killed countless women and infants in the past. But with this incredible progress has come a silent hazard that we are just beginning to understand. High rates of Cesarian sections and the overuse of antibiotics in mothers and newborns are altering the types of microbial species that mothers have always passed on to their newborns.

Microbes play a hidden role in the course of every pregnancy. For example, have you ever wondered why pregnant women gain more weight than can be accounted for by the size of their fetus and placenta? Bacteria are an answer.

The mother's blood carries nutrients, oxygen, and certain antibodies to the fetus via the permeable placenta. Fetal waste products and carbon dioxide are returned through the blood, and the mother's organs eliminate them. As far as we know, there are no bacteria normally present in the womb. It is believed to be a totally sterile environment, although this tenet of medicine is coming under question. However, we do know that particular infections, like rubella or syphilis, at such an early stage of life wreak havoc.

As the fetus grows, the mother's breasts and uterus start to enlarge. Simultaneously, and invisibly, the microbes in her intestinal tract begin to stir. During the first trimester, certain species of bacteria become overrepresented while others become less common. By the third trimester, just before the baby is born, even greater shifts occur. These changes, involving scores of species, are not random. The compositions change in the same direction across the dozens of women who have been studied. The pattern suggests that these microbes are up to something important, as if they are part of an adaptive trait designed to promote the pregnancy and prepare for birth.

A few years ago, Dr. Ruth Ley, a young scientist from Cornell who had just given birth, decided to study this process in her laboratory. One of the central biological problems of pregnancy is that the mother is responsible for feeding two people. She must find a way to shore up and mobilize energy and optimally divide it between her and her baby. Ruth hypothesized that the mother's gut microbes might help by reorganizing her metabolism in ways that benefit the fetus.

Ruth's team used germ-free mice to investigate the role of gut bacteria in pregnancy. Born and raised in sterile conditions, germ-free mice allow researchers to begin each experiment with a clean slate; the mice are free of all bacteria and, as far as we can determine, free of viruses and other microorganisms. They live in a plastic bubble. But scientists can end the germ-free state by introducing whichever microbes they want, one bug at a time, a few different ones, or entire communities from the contents of another mouse or even from a human. Prior work by many researchers has shown that human microbes will "take" in their new host and that these mice will accept this "graft." Such recipient mice are a kind of hybrid, mouse body and genes with a huge number of human microbes.

Ruth wanted to know what would happen if she took microbes from the intestines of pregnant women and put them into the intestines of germ-free mice. Her team compared two varieties of transplant: fecal microbes obtained from women during the first trimester of pregnancy and then from the third trimester. After inoculating the animals, she waited to see how they were growing. After just two weeks, the differences were substantial. The mice that received microbes from the third-trimester women gained more weight and had higher blood-sugar levels compared to the mice that received microbes from the first-trimester women.

If extended to humans, the experiment implies that many physiological and pathological features of pregnancy are controlled, at least in part, by the mother's resident microbes, which evolved to help her and themselves. When food is in short supply during pregnancy, as has often occurred in human history, the mother's microbes will shift their net metabolism so that more calories flow from food to her body. In this way, her microbes increase the odds that there will be a next generation, one that will provide a new home for them.

Thus shifts in microbial composition may be partially responsible for those extra pounds as well as for the increased sugar or glucose levels that commonly occur during pregnancy. It makes sense; mothers store more energy to optimize the success of their newborns.

One consequence of this process is that some women develop gestational diabetes; they can't handle the extra weight without stressing their systems. Most of the time, the problem is mild and resolves within weeks after delivery. Or, for an unfortunate few, the diabetes is severe. The good news stemming from Ruth's experiment is that one day we might be able to manipulate gut microbes in pregnant women to optimize their energy storage and tone down the diabetes. We might do this by restoring microbes harvested from the mother's first trimester, or maybe introduce microbes from women who don't develop diabetes. Or maybe give mothers prebiotics, foods tailored to nourish the composition of each woman's resident microbes. These studies open a world of new possibilities to make pregnancy a little safer.

■ ■ ■

As microbes in the mother's intestinal tract store up energy, another population of microbes—this time in her vagina—begins shifting as well. They, too, are preparing for the baby's birth. As noted earlier, women of reproductive age carry bacteria, primarily lactobacilli, which make the vaginal canal more acidic. This environment provides a hardy defense against dangerous bacteria that are sensitive to acid. Lactobacilli also have evolved a potent arsenal of molecules that inhibit or kill other bacteria.

During pregnancy, these tiger-mother lactobacilli flourish and predominate, crowding out other resident species and potential invaders. They are gearing up for the main event—birth—which occurs around the thirty-eighth or thirty-ninth week of most pregnancies. We don't know what initiates the process, why one woman is two weeks "early" while another is one week "late." My suspicion is that microbes are involved in this too.

When the mother's water breaks, a rush of fluid is unleashed into her vagina, sweeping up bacteria as it flows out of her body onto her thighs. This splash, now dominated by lactobacilli, rapidly colonizes the mother's skin. Meanwhile the baby is still in the womb preparing to exit. As labor progresses, contractions strengthen, forcing the cervix to fully dilate so the baby can emerge. A rush of hormones, including adrenaline and oxytocin, surge through mother and infant.

Whether the birth is fast or slow, the formerly germ-free baby soon comes into contact with the lactobacilli in the vagina. Very flexible, rather like a glove, the vagina covers the newborn's every surface, hugging its soft skin as it passes through. And with that hugging a transfer occurs. The baby's skin is a sponge, taking up the vaginal microbes rubbing against it. The baby's head faces down and is turned toward the mother's back to fit snugly in the birth canal. The first fluids the baby sucks in contain mom's microbes, including some fecal matter. Labor is not an antiseptic process, but it has been going on like this for a long time—at least 70 million years since our earliest mammalian ancestors.

Once born, the baby instinctively reaches his mouth, now full of lactobacilli, toward his mother's nipple and begins to suck. The birth process introduces lactobacilli to the first milk that goes into the baby. This interaction could not be more perfect. Lactobacilli and other lactic acid–producing bacteria break down lactose, the major sugar in milk, to make energy. The baby's first food is a form of milk called colostrum, which contains protective antibodies. The choreography of actions involving vagina, baby, mouth, nipple, and milk ensures that the founding bacteria in the newborn's intestinal tract include species that can digest milk for the baby. These species are also armed with their own antibiotics that inhibit competing and possibly more dangerous bacteria from colonizing the newborn's gut. The lactobacilli, which bloom in the mother's vagina as pregnancy's term nears, become the earliest organisms to dominate the infant's formerly sterile gastrointestinal tract; they are the foundation of the microbial populations that succeed them. The baby now has everything it needs to begin independent life.

Breast milk, when it comes in a few days later, provides the infant with further extraordinary benefits. It contains carbohydrates, called oligosaccharides, that babies cannot digest. Why would the milk contain energy-rich compounds that babies cannot use directly? The reason is microbes. The oligosaccharides can be eaten and used as an energy source by specific bacteria such as *Bifidobacterium infantis*, another foundational species in healthy babies. The breast milk is constituted to select for favored bacteria to give them a head start against competing bacteria. Breast milk also contains urea, a major waste product in urine, which is toxic to babies. Again, it is there to feed select beneficial bacteria by providing them with a source of nitrogen to make their own proteins in a way that does not cause the bacteria to directly compete with the baby for nitrogen. How clever of Nature to devise a system in which a maternal waste product is used to enhance the growth of bacteria beneficial for her babies.

Although babies are born into a world replete with diverse bacteria, the ones that colonize them are not accidental. In a continuation of the script that has evolved over eons, Nature selects for the good guys, the ones that provide the vital metabolic functions for the developing baby that nurture the cells lining the infant's intestines and crowd out bad guys.

Meanwhile, the mother's skin bacteria are busy colonizing her baby, and each kiss introduces her oral bacteria. Long ago, mothers used to lick their babies clean, and many animals still do that, transferring their microbes to the next generation. But now when a baby is born vaginally, everyone is in a hurry to clean it up, to remove the coating that covered it in the womb. This material, the vernix produced by fetal skin, has hundreds of useful constituents, including proteins that suppress specific dangerous bacteria. Because the hospital staff is in a hurry to get that baby nice and clean to present to Mom and for all of those photographs, the vernix is usually washed off. Are they doing that baby a service by washing off a coating that for eons probably protected all human babies? While no one yet has studied this in detail, my hunch is that the vernix serves to attract particularly beneficial bacteria and repel potential pathogens.

These first microbes colonizing the newborn begin a dynamic process, setting the stage for the subsequent more adultlike microbiota. They activate genes in the baby and build niches for future populations of microbes. Their very presence stimulates the gut to help develop immunity. We are born with innate immunity, a collection of proteins, cells, detergents, and junctions that guard our surfaces based on recognition of structures that are widely shared among classes of microbes. In contrast, we must develop adaptive immunity that will clearly distinguish self from non-self. Our early-life microbes are the first teachers in this process, instructing the developing immune system about what is dangerous and what is not.

As months go by, babies acquire more microbes from eating a more complex diet as well as from the people who surround them: mom and dad, grandmother, siblings, and other relatives and then later from neighbors, classmates, friends, and other humans. Eventually the process gets more random. The exposures differ, and the ones that stick differ. As discussed before, by age three, each of us has acquired our own unique foundation of microbes. To me, this is remarkable. In just three years, a great diversity of microbes self-organizes into a life-support system with the complexity of the adult microbiota. This occurs for each and every person. Those three years, when the first resident microbes are most dynamic, are when the baby is developing metabolically, immunologically, and neurologically. This critical period lays the foundation for all the biological processes that unfold in our childhood, adolescence, adulthood, and old age—unless something comes along to disrupt it.

■ ■ ■

Cesarian delivery is a largely unrecognized threat to the microbial handoff from mother to child. Instead of traveling down the birth canal picking up lactobacilli, the baby is surgically extracted from the womb through an incision in the abdominal wall. The procedure was invented in Roman times to save the baby's life. Mothers always died.

Today, C-sections are very safe, as they are almost always carried out in hospitals by experienced obstetricians. When the mother's or baby's life is in danger for any reason, emergency C-sections are performed, often with short notice. Prolonged labor or a failure for labor to progress, fetal distress, a ruptured amniotic sac or collapsed umbilical cord, high blood pressure in the mom, a breech position in the baby, or even a very large baby deemed too big to pass through the birth canal are common reasons for the surgery. In some populations, the rate for such emergency C-sections is pushing 20 percent, while in more holistic communities in Sweden the rate is about 4 percent.

C-sections are so safe that for a variety of reasons many women actually elect to have them. One is to diminish or avoid the pain of childbirth. This is not a trivial issue. For personal or cultural reasons, some women are frightened by giving birth. Given the availability of a safe alternative, it's a choice millions of women make. Some professional women choose the surgery because their work schedules matter. Some women schedule the surgery so they can be sure to attend an important wedding or graduation. And others, whose OB is in a group practice, opt for an elective C-section so they can be sure of having the doctor they want deliver their baby.

Physicians also influence their patients' choices of birth method. Some are very conservative when they see any signs of fetal stress or suspect that the mother will have problems. For example, when babies are in a breech position, natural childbirth can be dangerous. However, most breech babies turn head down not long before labor begins. On a more cynical note, it takes less time and fuss to do a C-section than to wait out a vaginal birth. And most doctors and hospitals make more money from performing operations like C-sections than from natural births.

For all of these reasons, the U.S. C-section rates increased from fewer than one in five births in 1996 to one in three births in 2011—a 50 percent increase. If this trend continues, half of all U.S. babies (2 million a year) might be delivered surgically by 2020.

C-section rates around the world show astonishing variation. In Brazil, over 46 percent of all births are by C-section. In Italy, it is 38 percent, but in Rome, where the operation is thought to have been invented, the rate is 80 percent. In Scandinavian countries, which pride themselves on medical conservatism, fewer than 17 percent of births are Cesarian, and the rate is 13 percent in the Netherlands.

Why such differences? The act of giving birth is the same everywhere. The only explanation is variations between local practices and customs. Women from Rome, who these days tend to have but one child, often become pregnant in their thirties and often have busy careers. They are twice as likely to have a C-section than women in the rest of Italy, suggesting that the procedures are not driven by pelvic anatomy.

But so what? Do high C-section rates matter? Why not perform a Cesarian if it makes the mother more comfortable and is easier on the physician, if there is no cost other than the hospital bill?

Well, there is a cost—a biological one—and it affects the baby. A few years ago, my wife, Gloria, was stuck for a couple of weeks in Puerto Ayacucho, the capital of Amazonas State in Venezuela. She had been conducting nutritional and microbiological studies there for nearly twenty years and had permits to sample the microbiomes of Amerindians living there. She had been waiting to go into the jungle to collect microbes from a newly discovered Amerindian village, but the helicopter flight assigned to the health team was canceled. So, thinking she might make herself useful, she headed over to the local hospital. Would the microbes found on newborn babies delivered vaginally or by C-section vary in any way? No one had ever conducted this kind of study.

Nine women, aged twenty-one to thirty-three years, and their ten newborns participated. Four mothers delivered naturally and five had C-sections that had been planned. Gloria sampled each mom's skin, mouth, and vaginal microbes one hour before delivery. By DNA sequencing, she showed that the women all had similar proportions of the major bacterial groups present at each site.

Each baby's skin, mouth, and nose were sampled less than fifteen minutes after birth. She sampled their first stool, called meconium, twenty-four hours later.

While all of the mothers had many different types of bacteria on and in their bodies before giving birth, the moms who gave birth vaginally now had the signature splash of amniotic fluid on their skin with lots of lactobacilli. Most important, the babies showed a different pattern, depending on mode of delivery. The mouths, skin, and first bowel movements of babies born vaginally were populated by their mother's vaginal microbes: *Lactobacillus, Prevotella,* or *Sneathia* species. Those born by C-section harbored bacterial communities found on skin, dominated by *Staphylococcus, Corynebacterium,* and *Propionibacterium.* In other words, their founding microbes bore no relationship to their mother's vagina or any vagina. At all the sites—mouth, skin, gut— their microbes resembled the pattern on human skin and organisms floating in the air in the surgery room, including those on the skin of nurses and doctors and bacteria on sheets from the laundry. They were not colonized by their mother's lactobacilli. The fancy names of these bacteria don't matter as much as the notion that the founding populations of microbes found on C-section infants are not those selected by hundreds of thousands of years of human evolution or even longer.

Gloria studied newborns, but we know from other researchers that as babies are exposed to the wider world during the first months of life the microbiomes of C-section and vaginally delivered infants begin to converge. The earlier differences between them diminish. One reason may be that sooner or later everyone gets exposed to organisms that play similar roles in the body. But maybe those initial differences at birth are more important than we have realized. What if those first microbial residents provide signals that critically interact with cells in the rapidly developing baby's body? We will consider this in later chapters.

■ ■ ■

Another threat to a baby's newly acquired resident microbes involves antibiotics given to the mother. After thalidomide, the medical community became much more cautious about giving drugs to pregnant women. Does that mean that the antibiotics recommended for pregnant women are safe? And who are they safe for, the mother or her fetus?

Most doctors consider it safe to prescribe penicillins, including ampicillin, amoxicillin, and Augmentin, for all sorts of mild infections in pregnancy—coughs, sore throats, urinary tract infections. Sometimes when doctors think that the mother has a viral infection they also give antibiotics as well "just to be sure" (for the small chance that it is actually a bacterial infection). As we know, the antibiotics affect the mother's resident microbes in all locations, inhibiting susceptible bacteria and selecting for resistance. The closer the dose is to birth, the greater the possibility that she will pass a skewed population of microbes to her baby.

Then comes the birth itself. Women in labor routinely get antibiotics to ward off infection after a C-section and to prevent an infection called Group B strep. About 40 percent of women in the United States today get antibiotics during delivery, which means some 40 percent of newborn infants are exposed to the drugs just as they are acquiring their microbes.

Thirty years ago, 2 percent of women developed infection after C-section. This was unacceptable, so now 100 percent get antibiotics as a preventive prior to the first incision.

Antibiotics are also used to prevent a serious infection in newborns caused by Group B strep. This bacterium lives in the gut, mouth, skin, and sometimes the vagina and rarely causes any problem in the mother. Recall that streptococci are among the most common groups of microbes found in the human body. Between a quarter and a third of pregnant women in the United States carry Group B strep.

But sometimes Group B strep can be lethal to newborn babies whose immune system is not up and running. While such infections are uncommon, professional groups recommend that all pregnant women be screened for the microbe near the time of delivery. If they are positive, they get a dose of penicillin or a similarly effective antibiotic shortly before the baby descends the birth canal.

But the problem, of course, is that we know antibiotics are broad in their effects, not targeted. While the antibiotic kills Group B strep, it also affects other often-friendly bacteria, killing susceptible bacteria and thus selecting for resistant ones. This practice is altering the composition of the mother's microbes in all compartments of her body just before the intergenerational transfer is slated to begin.

The baby also is affected in similar unintended ways. Any antibiotic that gets into the bloodstream of the fetus or into the mother's milk will inevitably influence the composition of the baby's resident microbes. A baby who starts life with penicillin in its blood or its gut is different from one without because of how the antibiotic affects each child's developing microbiome, but we are only beginning to understand what this means. One likely scenario is that the antibiotics reduce some taxa of bacteria and enhance others. Whether the effect is transient and trivial or the first step in a cumulative process is unknown. I believe this is an important area for further study.

All in all, each year in the United States well over a million pregnant women are Group B strep–positive, and all will get intravenous penicillin during labor to prevent their babies from acquiring Group B strep. But only 1 in 200 babies actually gets ill from the Group B strep acquired from his or her mother. To protect 1 child, we are exposing 199 others to antibiotics. There must be a better way.

When penicillin had no perceived cost other than occasional allergies, massive overtreatment did not seem like a problem. But what if changing microbial compositions affect the baby's metabolic, immunologic, and/or cognitive development? As we will see from experiments that my lab and others have conducted, such fears have a real basis.

Another important consideration is that while fewer babies today are born with serious Group B strep infections, the rate of other infections may be rising. By killing or inhibiting some bacteria, penicillin selects for other, resistant bacteria, such as certain virulent strains of E. coli, which themselves can infect susceptible newborns. It is possible that, in terms of avoiding serious neonatal infections, the net positive effect of exposing a million mothers each year to this penicillin is lower than anticipated. Also frightening is a conversation I recently had with a colleague who told me that his wife tested negative for Group B strep, but the doctor still wanted to treat her with high-dose penicillin (in case they had "missed" it). Fortunately, she refused.

Many women get yet another dose of antibiotics when they have an episiotomy, a surgical cut of the vaginal wall to prevent tearing and excessive bleeding as the baby's head crowns. A generation ago, half of American women giving birth received episiotomies. Now, due to changing custom, it is a third. But in Latin America, nine of ten women giving birth vaginally for the first time get the procedure. The rates vary according to local custom and physician advice. But most mothers never realize that they received antibiotics when their babies were born; either they were never told or the news didn't register.

Finally, the babies are directly exposed. Most parents are not aware that all American-born neonates today are given an antibiotic immediately after birth. The reason is that many years ago, before there were antibiotics, women who had gonorrhea, a sexually transmitted infection, were unable to clear the causative bacteria, even though they had no symptoms of the disease. Their infection would be discovered only when their baby developed a terrible eye infection. As the babies passed through the birth canal, they became inoculated on their face. Sometimes, the eye infection that ensued, called gonococcal ophthalmitis, was so severe that it blinded the baby.

For more than a hundred years, babies have been given eye drops to prevent this infection, first silver nitrate and then, more recently, antibiotics. Although much of the antibacterial effect remains local, the broad-spectrum antibiotics are absorbed into the bloodstream and circulate throughout the newborn's body. The dose is low but likely is affecting the composition of the infant's resident microbes just when the founding populations are developing. My lab hopes to soon start a study to measure the magnitude of the perturbation.

So 4 million babies born in the United States every year are being treated to prevent an illness that although catastrophic occurs very rarely. We should be able to develop a better way to screen, so we can target those babies at the highest risk, perhaps a few hundred among the millions of births a year. In Sweden, babies do not receive silver nitrate or antibiotic eye drops, with no effect on rates of infection, so there is precedent for a much more careful assessment of risk and benefit. But the public-health formulas to treat babies by the million to protect hundreds at risk were all based on the notion of essentially no biologic cost for the antibiotics given. What if it is not free?

9.

A FORGOTTEN WORLD

The continuing overuse of antibiotics in children and adults, changing birth practices, and the dosing of our farm animals with mountains of drugs inevitably have an effect on our bacteria, friend and foe alike. More than fifteen years ago, I began to think about what those effects might look like and to formulate the idea that the loss of our ancient, functionally conserved microbial inhabitants has led to the modern plagues I have mentioned: obesity, juvenile diabetes, asthma, and the rest.

The next five chapters explain the results of experiments I performed in my laboratory, first at Vanderbilt University and since 2000 at New York University, in an effort to confirm this hypothesis. The work has had many unexpected twists and turns, failures and successes, lots of hard work, and disappointments of every type. Still and all, the work is ongoing; the exciting days more than equal the busts, and we have been getting somewhere. Some days the results are so clear and so beautiful (thanks to great students who have learned to present their findings with artistry) that I can't believe they are quite true. But the good ones keep happening again and again, and that's how we know they are real. And I am running as fast as I can.

The ancient stomach bacteria *Helicobacter pylori* have been my guide for nearly thirty years. When they were discovered, or as you'll soon see rediscovered, in 1979, their impact on human health was not obvious. Only later did it became clear that they led to specific diseases. But for the past eighteen years, my research has focused on how *H. pylori* keeps us healthy.

Making us sick/keeping us healthy—this may seem contradictory, but this dual nature has plenty of company in the natural world. More than fifty years ago the microbial ecologist Theodore Rosebury coined the term *amphibiosis*, the condition in which two life-forms create relationships that are either symbiotic or parasitic, depending on context. One day the organism is good for you—let's say it fights off invaders. The next day it turns against you. Or, on any day, both happen simultaneously. Our colonization by the viridans streptococci discussed earlier is an example. Amphibiosis is all around us, including in our work relationships and marriages. It is at the heart of biology, in which the constancy of natural selection forces myriad nuanced interactions.

Amphibiosis is a more precise term than *commensalism*. Commensalism has been used to describe guests who come to the dinner table to eat; it's not so hard to serve them an extra meal, but they don't contribute much if anything to the upkeep of the kitchen. Until recently, that is more or less how we considered the microbes living in the human body, what we called the normal flora. Now we know that Rosebury's amphibiosis better describes the more complex relationship between our bodies and our indigenous organisms. *Helicobacter pylori* is the best model I know for these interactions, and exploring its biological interface with humans can help us understand the wider world of our normal resident microbes.

H. pylori are curved bacteria found in essentially only one place: the human stomach. Billions of them live in a thick layer of protective mucus just inside the stomach wall. Mucus lines the gastrointestinal tract from your nose to your anus. It is a gel that helps food slide down and protects the walls of the GI tract from the digestive processes. In each part of the GI tract the locally produced mucus differs in its chemical composition, and, importantly, each zone has its own bacterial species. Your stomach mucus is particularly thick, forming a barrier against the highly acidic environment needed to break down food and repel pathogens. It's here that we find *H. pylori*.

Helicobacter pylori have deep roots in evolution. The first primitive mammalian ancestor had a single stomach that laid the blueprint for all the stomachs that followed. As mice, monkeys, zebras, and dolphins radiated in separate directions, so did their stomachs, each with its own acid secretions, mucous layer, and the microbes evolved for that niche. Today we can recognize many *Helicobacter* species in mammals: *H. suis* in swine, *H. acinonyx* in cheetahs, *H. cetorum* in dolphins, and *H. pylori* in humans.

We know from genetic studies that humans have carried *H. pylori* for at least 100,000 years, which is as far back as we can determine using available methods. It's reasonable to assume that we have had the microbe with us since the origin of *Homo sapiens* about 200,000 years ago in Africa. It's been a long-term relationship, not a one-night stand.

Genetic analyses also tell us that all modern *H. pylori* populations derive from five ancestral populations: two from Africa, two strongly associated with Eurasia, and one with East Asia. We can trace the movement of *H. pylori* as people migrated around the world, carrying the organism as hidden passengers in their stomachs. Studies from my lab provide evidence that when humans crossed the Bering Strait from the Old World into the New World approximately eleven thousand years ago, they had East Asian strains of *H. pylori* in their stomachs. Nowadays, European strains predominate in South America's coastal cities as a result of the racial mixing that occurred after the Spanish arrived. But pure East Asian strains can still be found among Amerindians living deep in the continent's jungles and highlands.

Until recently *H. pylori* colonized virtually all children early in life, shaping the stomach's immune responses in ways favorable to the microbes as well as to the child. Once *H. pylori* take hold, they are remarkably persistent. Many other microbes that we come into contact with, including bacteria in your dog's mouth, bacteria in yogurt, and viruses that cause the common cold, are not. They pass through us transiently. But *H. pylori* have evolved a strategy for hanging on even as some of them are swept out of the body by peristalsis, the motion that pushes mucus, food, and wastes along your gastrointestinal tract and out your bottom. *H. pylori* can swim and multiply sufficiently rapidly to maintain their numbers for most of a person's life. For millennia, these bacteria have fought successfully against the tide and until recently absolutely dominated the stomach. But nothing prepared *H. pylori* for the twentieth century, which is the setting of my main story. But first we must go back a little further in time.

■ ■ ■

In the nineteenth century, early pathologists used microscopes to compare normal and abnormal tissues in people who were ill. This was the beginning of the medical discipline of pathology, and they immediately saw differences. Normal tissues have regular shapes and great symmetry, lines and lines of cells in perfect rows. But infected tissues, such as a wound, an inflamed joint, or a swollen appendix, are infiltrated with white blood cells that sometimes form sheets, like an endless army of soldiers. Other times, the white blood cells form a rim around a space filled with pus, which contains remnants of tissue destroyed in the battle between the white blood cells and a pathogen.

Such infiltrations, called inflammation, correlate with the swelling, redness, heat, and tenderness that we experience with an infection or arthritis. Sometimes the inflammation is extensive, as in a raging abscess. Or it can be subtle, as in a muscle that has been overexercised and is sore the next day.

Those early pathologists and clinicians also looked in the stomach, where they observed, in essentially everyone, large numbers of bacteria that were curved like commas or were S-shaped like spirals. But these organisms were very particular in their growth requirements and could not be isolated in the kinds of cultures established by microbiologists on petri dishes. Because these organisms could never be grown in the lab, as could many other organisms in the gastrointestinal tract, their identity remained unknown and as a result they were ignored. They were deemed to be just some ordinary commensals that everyone shared, not long after which they were forgotten.

Within a few decades, physicians were being taught that the stomach is sterile and completely free of bacteria. Of course there had to be a reason why the stomach, which is next door to the bacteria-rich intestine, had no bacteria in it. And, having forgotten all about the curved bacteria, the professors invented a reason: obviously nothing could survive in the highly acidic stomach. Since stomach acid is similar in strength to the acid found in a car battery, it made sense to deduce that bacteria could not live in that environment. Our view of the bacterial world was then pretty limited; we had no idea that bacteria can thrive in volcanoes, hot springs, granite, deep-sea vents, and salt flats.

Doctors also knew that a highly acidic stomach can cause trouble. It can become injured and inflamed, and when that becomes particularly intense, the surface of the stomach wall can break, forming an ulcer. Ulcers, which also form in the duodenum, the first part of the small intestine just downstream from the stomach, can cause severe pain. They can erode into a blood vessel, causing substantial bleeding that sometimes is fatal. Or they may burrow through the stomach wall, creating a perforation connecting the stomach interior to the usually sterile space called the peritoneum. In the old days, that almost always was fatal as well. Between meals or in the middle of the night, people with ulcers can experience a gnawing or burning pain in their abdomen or have bloating or nausea. These ulcers can persist, or they can come and go.

In 1910 a German physiologist, Dragutin Schwarz, recognized that for an ulcer to occur, the stomach must contain acid. Elderly people whose stomach acid had naturally dissipated never got ulcers. Schwarz's dictum was no acid, no ulcer. So doctors figured that the way to treat ulcers was to reduce stomach acidity. Generations of patients were advised to drink milk, take anti-acids, or undergo surgery that eliminated or reduced the stomach's ability to produce acid. Moreover, stress seemed to make ulcers worse, which would explain why they waxed and waned. People were urged to control their stress along with their stomach acid. In fact, as a medical student I learned that men with ulcers had trouble getting along with their mothers and that ulcers were one of the best examples of psychosomatic illness. This particular lecture was given by a prominent psychiatrist whose ulcer treatment involved psychotherapy. Not surprisingly, each of the many popular remedies had important limitations, and peptic ulcer disease, as it came to be called, remained a major problem.

Then in 1979 Dr. Robin Warren, a pathologist in Perth, Australia, again noticed bacteria present in the mucous lining of the stomach. Using routine and later specialized stains, he could clearly see the comma and S-shaped bacteria. He further noted that stomach walls of people with bacteria showed signs of inflammation under the microscope or what pathologists like Warren typically call gastritis. Nearly a century after the initial discovery of bacteria in the stomach, Warren realized that the stomach was not sterile after all. It contained bacteria, and he correctly deduced that they must be involved somehow in the inflammation. But what kind of bacteria were they? Why didn't stomach acid kill them off?

Within a few years, Warren shared his observations with Dr. Barry Marshall, a young trainee who had an "aha moment" of his own. He learned from reading the medical literature that almost everyone who has peptic ulcer disease also has gastritis. If the bacteria were related to gastritis, he reasoned, they also might be related to ulcers. They might even cause these peptic ulcers.

The two researchers studied biopsies from patients with and without ulcers. Nearly everyone with an ulcer had both the S-shaped bacteria and gastritis. But many people without an ulcer also had gastritis and bacteria. They concluded that the mysterious bacteria might be necessary but not sufficient to cause ulcers, just as in the case of gastric acidity.

Doctors (me included) were taught that gastritis is a pathological inflammation of the stomach. But hindsight allows me to question whether it really is pathological or instead a normal condition of the stomach in reaction to coexistence with bacteria. We'll come back soon to this distinction, which is not just academic but is in fact central to understanding our relationship with *H. pylori*.

In April 1982, using methods developed over the previous few years to isolate *Campylobacter* organisms from fecal specimens, Warren and Marshall cultured the S-shaped gastric bacteria for the first time. They accomplished the feat that had eluded German, Dutch, and Japanese scientists nearly a century earlier. As noted in chapter I, they first called these bacteria "gastric campylobacter-like organisms" (GCLO), then *Campylobacter pyloridis*, then *Campylobacter pylori*. Several years later, after more extensive study, it became clear that these organisms were not campylobacters at all but previously unknown cousins. That's when they received their new name: *Helicobacter pylori*. Within months of Warren and Marshall's first publication in 1983 in the *Lancet*, other investigators began finding these "new" organisms in the stomach and reporting their association with gastritis.

But Marshall wanted proof that these organisms could play a causal role in ulcers and were not just passengers. So in 1984 he used himself as a guinea pig. After testing showed that his stomach was *H. pylori*–free, he swallowed a culture of the organisms. At first nothing happened. But after a few days he developed indigestion. A new biopsy of his stomach revealed the presence of *H. pylori*, but even more important he had gastritis; his stomach hurt, and he had bad breath.

A few days later, a second biopsy showed that the gastritis was largely gone. But because Marshall worried that the organism might persist, he took a single antimicrobial agent, tinidazole, and, as far as what has been published, he was never bothered again by *H. pylori*.

Marshall's self-experiment showed that *H. pylori* caused the gastritis rather than merely thriving in an environment created by it. But his acute gastritis lasted only a few days before getting better on its own. His condition was different from the usual chronic gastritis, which is present for decades in people with *H. pylori* in their stomachs. Moreover, Marshall took an antibiotic that we now know is ineffective in clearing *H. pylori* when taken alone. So with the benefit of hindsight we know that the infection and inflammation had spontaneously cleared. Most important, Marshall never developed an ulcer.

Nevertheless this dramatic experiment convinced most skeptics to accept the idea that this common organism was indeed a pathogen. Since *H. pylori* caused inflammation, it was obviously a bad microbe. Most people remember the experiment as one in which a crazy but brave Australian drank bacteria and caused an ulcer, thus proving his theory. Of course that is incorrect, but it caught the world's attention.

Next, to see whether *H. pylori* might have a direct role in causing ulcers or might be just bystanders, Marshall and Warren treated ulcer patients with regimens containing bismuth, an antibacterial agent, or with regimens without bismuth. The results were clear: the rate of ulcer recurrence was much lower in the patients who received bismuth. And other investigators found the same relationships in their own studies.

Doctors could now treat ulcer patients with antibacterial agents, including antibiotics. This was revolutionary. Ulcers could be cured. Good-bye to the idea that stress caused ulcers; hello to microbes.

For their isolation of *H. pylori* in pure culture, for establishing its association with gastritis and with peptic ulcer disease, and for changing the treatment of ulcer disease, Marshall and Warren were awarded the Nobel Prize in Physiology or Medicine in 2005. This recognition solidified the notion that *H. pylori* was a major human pathogen and that anyone who had it their stomach would be better off without it.

But many mysteries about ulcer disease remained. Why does it affect men so much more than women, although they carry *H. pylori* at about the same frequency? Even though *H. pylori* is carried from early childhood to old age, why does ulcer disease start to appear in the third decade of life, peak over the next twenty years or so, and then decline? Why does an ulcer form and then heal after a few days or weeks and then recur weeks, months, or years later? Finding the link with *H. pylori* enabled us to better treat ulcers and to prevent their recurrence, yet we still understood little about the biology of the disease.

■ ■ ■

Having heard the first presentation of Warren and Marshall's work at the International Campylobacter Workshop in Brussels in 1983, I was initially skeptical, in particular about Marshall's assertions. Clearly they had discovered a new microbe, but Marshall's claims about ulcers were not convincingly supported by the evidence he presented. Yet as Marshall and others in the field kept showing relationships between the organism and gastritis and ulcers, I decided that my lab should get involved. In 1985, we began to study the organisms themselves (still called campylobacters) and found that they were diverse, but that people who had them in their stomachs formed antibodies to them in their blood.

In 1987 my longtime collaborator Guillermo Pérez-Pérez and I developed the first blood test to accurately identify carriers of *H. pylori* based on their having antibodies to the organism. Like most scientists, we wanted to know our own status. One of the first things we discovered is that I was positive. I must admit, I was surprised. Like most people in the world who have *H. pylori* in their stomachs, I had no symptoms. My belly felt fine, although when I learned the test results I began to feel a little queasy. But the test opened many windows for us. We could obtain blood specimens from people of all ages all over the world, with different kinds of diseases or none, and, with our test, determine who had the organisms hidden in their stomachs, so we could look for relationships with various illnesses.

I wanted to know why only some of the people with the microbe developed ulcers. We had shown that *H. pylori* strains varied considerably, but we did not know whether these differences would determine whether or not a particular strain would cause disease. For example, nearly all of us carry *E. coli*, which is mostly harmless. Only a few types are very dangerous because they carry genes that code for particular proteins, called virulence factors, that make us sick. We wondered whether any *H. pylori* strains had virulence factors. Could such differences explain who got sick and who did not? Was the observed diversity clinically relevant?

After two years of study, we identified a protein in *H. pylori* that fit the bill. It was essentially always present in the strains found in people with ulcers. People without ulcers had it about 60 percent of the time. So while it seemed necessary for ulcers, it was not sufficient. Still it was a very good lead. Could we find the gene that encoded this protein? In 1989 we created a "library" of *H. pylori* genes within *E. coli* cells. This simply means that we could use the *E. coli* cells as microscopic factories to produce *H. pylori* proteins. Each cell churned out only one or two of the estimated 1,600 *H. pylori* proteins. Then we took the blood serum of a person who tested positive for the microbes (again it was me) and screened the library to see if any of the *E. coli* cells produced any proteins recognized by my antibodies. In other words, we went fishing, and we landed a big fish. The very first clone that my serum recognized coded for the same protein that we had associated with ulcers. We named it *cagA*, for cytotoxin-associated gene.

Later we learned how clever these microbes are. These virulent strains contain a cluster of genes that not only make highly interactive proteins, such as CagA, but also form a system for injecting these materials from the bacterial cells into host cells. This meant that my *H. pylori* cells were churning out the CagA protein and constantly injecting it into the cells of my stomach wall. This revved up inflammation, which as far as I was concerned at that time was not a good thing.

A second finding we made at that time was that all *H. pylori* strains possess a protein that in sufficient quantity pokes holes in the epithelial cells that line the stomach wall. Some strains make bigger holes than others by secreting a protein that we discovered and named VacA.

■ ■ ■

After studying Marshall and Warren's work showing that *H. pylori* played a role in ulcer disease and gastritis, we had another relationship on our minds: whether the microbe was associated with stomach cancer. Cancer is the main scourge of the human stomach. It is an awful disease. After diagnosis, a person has less than a 10 percent chance of being alive five years later. In 1900 stomach cancer was the leading cause of cancer death in the United States. It still is the number-two cause of cancer death in the world, after lung cancer.

In 1987 we tried to convince the National Cancer Institute to work with us on a possible relationship between *H. pylori* and stomach cancer; however, we were turned down. But two years later I was contacted by Dr. Abraham Nomura, principal investigator of the Japan-Hawaii Cancer Study based in Honolulu. He and his colleagues had done pioneering work on the disease risks of Japanese-Americans living in Hawaii, and he wanted to use our blood test to study the risk for stomach cancer related to *H. pylori*. I jumped at the chance.

Between 1965 and 1968, more than 7,400 Japanese-American men born between 1900 and 1919 enrolled in the Honolulu Heart Study. These men, veterans of the 442nd Regiment who fought in the U.S. Army with great distinction during World War II, were heroes of mine ever since I had learned about them from reading James Michener's book *Hawaii* as a boy. When Japanese-Americans were being rounded up and incarcerated on the West Coast of the United States, these men risked (and some lost) their lives and limbs to defend their country. The late Senator Daniel Inouye was one of them.

By 1989, blood samples from nearly 6,000 of these veterans had been obtained and frozen. During the interim, more than 137 men developed stomach cancer, and of these 109 could be studied. We matched them with 109 men who did not develop stomach cancer and examined their blood for antibodies to *H. pylori*. One strength of the study is that the blood specimens were obtained an average of more than twelve years before the cancer had been diagnosed. This window of time could help establish a causal relationship.

We asked two simple questions: who had *H. pylori* in their stomach in the 1960s, and did having the organism relate to getting cancer later on?

Our findings were dramatic. We discovered that those who carried *H. pylori* back then were six times more likely to develop stomach cancer over the next twenty-one years than those who were negative. I presented this finding as a "late breaker" at the same conference where Marshall had presented his findings about ulcers eight years earlier. Other parallel studies conducted in California and in England yielded similar results. Later we found that those who had the *cagA*-positive type of strains had double the risk.

It soon became clear that *H. pylori* was not just along for the ride. Carrying *H. pylori* preceded the development of stomach cancer. In 1994, based on our work and that of others, the World Health Organization declared *H. pylori* a Class I carcinogen for its relationship with stomach cancer. It was like smoking and lung cancer: no arguing about cause and effect.

No wonder doctors around the world began to believe that "the only good *Helicobacter pylori* is a dead one." From ulcers to cancer, everything indicated that carrying *H. pylori* is costly to humans. Doctors everywhere started to look for it in patients who had any kind of gastrointestinal symptoms; and if they found it, they would eliminate it using antibiotic-based treatment regimens. Part of the rationale was the fear of the cancer, and part of it was to treat the symptoms that patients had. But except for ulcers, clinical trials did not show that symptoms improved any more than by chance alone. Still, everyone was happy to eliminate *H. pylori* whenever they found it.

■ ■ ■

Yet for years I kept returning to a question: Why did Warren discover the association of *H. pylori* with gastritis when it had been missed for so long? Eventually I remembered learning that nineteenth-century pathologists had found those curved and spiral organisms in the stomach of virtually everyone. By the 1970s, in the slice of Australia where Warren worked, only about half of the adults were positive. Pathologists in other developed countries saw the same thing: *H. pylori* and associated gastritis in a fraction of the people, not everyone.

However, in contemporary studies from Africa, Asia, and Latin America, nearly all adults carried *H. pylori*. It was as if they had nineteenth-century stomachs, while we "developed peoples" had twentieth-century stomachs.

I made a leap: Warren was able to find the association with gastritis because *H. pylori* was no longer universal; it was disappearing. This ancient organism was becoming extinct. Other researchers noticed that *H. pylori* was less common in younger people, but they all thought it was a sign of progress, and of course in a way it was.

Our more recent work shows that most people born in the United States in the early twentieth century carried the organism. But fewer than 6 percent of children born after 1995 have it in their stomachs. Similar trends have been documented in Germany and Scandinavia. In fact, wherever we look *H. pylori* is disappearing from humans, most rapidly in developed countries but also in developing areas. This variation is not based on geography but rather on socioeconomic status. Poor people tend to have *H. pylori*; wealthier people tend not to. We see this everywhere we look all over the world. The presumption has been that it's better to be without *H. pylori* just as it is to be wealthier.

But why is *H. pylori* disappearing? Why is an organism that has survived so long in nearly all of our ancestors as the dominant bacteria in our stomachs been receding everywhere we look? The answer can be summarized in two words: modern life. A persistent colonizer like *H. pylori* must deal with two major biological problems: how it is transmitted to new hosts and how it is maintained until passed forward.

Transmission is the biggest bottleneck. *H. pylori* lives only in humans. As noted earlier, we do not get it from our pets, farm animals, or foods of animal origin, as we get other transient organisms such as *Salmonella*; nor do we get it from dirt. The major reservoir in the world for *H. pylori* is the human stomach. The microbe must pass from one stomach to another, and the only way to do that is to go either up or down the gastrointestinal tract.

H. pylori can easily travel up from the stomach to the mouth via burping or reflux. It can set up shop in dental plaque. In many parts of the world, mothers prechew food and pass it on to their babies' mouths, thereby transmitting the microbe. When people vomit, *H. pylori* is present and can be carried by the air for several feet, contaminating the nearby environment—comforting thought.

Down is easier. Everything in the gastrointestinal tract has the potential to come out at the bottom in the feces, and both *H. pylori* DNA and live organisms have been detected there. Usually live *H. pylori* are excreted at very low levels, but more come out after a microbial bloom. When hygiene is bad, as has been the case most of the time we humans have been on the planet, feces contaminate food and water. Fecal *H. pylori* is thus transmitted to the next person.

Young children are the most susceptible to *H. pylori*. They seem to resist it in their first year of life, but after that, in countries where sanitation and hygiene are poor, about 20–30 percent acquire it every year. Between the ages of five and ten, most kids get colonized, often with several different strains. After that, the transmission frequency drops.

Why the decline over the past one hundred years? One obvious reason is sanitation. In the late nineteenth century, cities began to provide clean water for their citizens from watersheds that were not grossly contaminated with feces and with the important advance of chlorination. Such measures helped prevent the transmission of cholera, typhoid fever, hepatitis, and childhood diarrheal illnesses. Resounding public-health successes, they account for a major part of our improved health and longevity in the first half of the twentieth century. Yet in preventing the spread of pathogens, these practices also reduced transmission of our ancient colonizing microbes, like *H. pylori*. The benefit of clean water is so huge that we must not denigrate its importance, but we should also recognize the potential for hidden consequences that diminish our ancestral microbiome.

Drinking contaminated water is how a child could acquire *H. pylori* from a stranger, but most transmission occurs closer to home. As indicated above, a baby can get *H. pylori* from her mother chewing his or her food. We don't know all the ways that children get *H. pylori* from their mothers, but studies have shown that the number-one predictor of whether a child has *H. pylori* is whether his mother has it.

Children also get *H. pylori* and other microbes from their older siblings. In a sense, the siblings amplify the transmission from mom, providing new opportunities for the organism to spread. Big families are an important reservoir for the organism, yet in developed countries families have been getting smaller. In a family with five children, 80 percent of kids have an older sibling. With two children, it is 50 percent. With an only child, it is zero. Before people became more prosperous, kids used to sleep in the same bed, sometimes with parents as well. Such close contact facilitated transmission of microbes, especially during critical windows, like early childhood.

Interestingly, when adults live together, as we showed in two studies, the risk of transmission of *H. pylori* to one another seems pretty low. We studied couples who were going to an infertility clinic, a group that might be having more physical contact than others; positivity in one member was no more likely to be associated with the same status in the other than by chance. We also looked at adults who came to a clinic for sexually transmitted diseases. With many organisms, such as the ones causing gonorrhea and syphilis, the more sexual partners you have, the more likely you are to acquire them. Not so with *H. pylori*; the organism hardly ever spreads from adult to adult.

If it is actually acquired in childhood, *H. pylori* must be maintained, so it can be transmitted to the next human generation. We know from both human and monkey experiments that the organism needs a period of time to adjust to its host. Some don't make it, as in the case of Barry Marshall's self-inoculation. If conditions are difficult for the organism, the success rate of transmission goes down.

Given the number of doses of antibiotics given to our children today, it's easy to imagine that a major impact on *H. pylori* colonization comes from treating all those sore throats and earaches. A single course of an antibiotic will eliminate the microbe in 20–50 percent of patients. When children are given those same antibiotics, they stand a similar chance of losing their *H. pylori*.

It is my belief that each time they take a course of antibiotics, and with each course given in a population of children, a few more kids lose the organism. Across the whole population, the trend is cumulative. This practice is a paradigm for the disappearance of other of our ancient organisms. Fitness is not guaranteed. In its protected gastric niche for eons, *H. pylori* was not at all prepared for the onslaught of antibiotics in the last seventy years.

The loss is multigenerational. Studies show that if the mother has lost her *H. pylori*, chances are small that her child will acquire it. And so it will go, generation after generation. Starting with sulfa drugs in the 1930s and then penicillin and others in the 1940s, in the U.S. and western Europe we already are in the fourth or fifth generation of antibiotic users. Remember the recent data implying that young people have had about seventeen courses of antibiotics by the age of twenty, essentially when the women are starting their child-bearing years. And loss of *H. pylori* in an older sibling removes another opportunity for transmission. Clean water, smaller families, and lots of antibiotics create a triple whammy against *H. pylori*.

A final cause of its disappearance is that *H. pylori* like to have sex with other *H. pylori*. This is an essential part of their biology. Some bacteria are more reclusive, like the ones that cause anthrax or tuberculosis. For *H. pylori*, free love is a way of life. In the old days, the average person probably had several different *H. pylori* strains in his or her stomach, just as we see today in people in developing countries. Contaminated water is again part of the reason. These mixtures of *H. pylori* strains represent a robust community. With their constant exchange of genes with one another, their populations shift, reflecting the changing tides in the stomach. Such gene exchange makes the community very adaptable, so it can take advantage of all of the resources that a stomach can provide. The overall community can be sustained for years, even decades. This is the strategy that *H. pylori* have evolved over the millennia: organisms competing with one another as always but also cooperating to ensure transmission to a new host. But in recent years, as transmission and maintenance have become more and more difficult, the number of individual strains able to colonize the average stomach has declined from three to two to one to zero.

■ ■ ■

As I came to realize that in just a few generations the microbial ecology of the human stomach has changed markedly, I wavered from the conviction that *H. pylori* are only bad. I could see that while *H. pylori* caused inflammation, it had been with us for a very long time and that most people who became ill, especially with stomach cancer, were much older. The average patients were in their seventies, and the cancer rates were higher still among people in their eighties. Across our entire population, the cost of *H. pylori* was not as high as that for malaria or diphtheria, for example, which kill children.

I began to think that maybe under some circumstances the inflammation caused by *H. pylori* could be good for us. My original ideas were fuzzy; I didn't know what good there could be. I only knew that when ancient dominant organisms disappear, there are bound to be consequences. This was heresy to most of my colleagues; having discovered *H. pylori* as a pathogen, they focused on the costs and considered it imperative to speed up its departure from this planet. They were not thinking about amphibiosis, just elimination.

Later we did find those benefits. In retrospect they seem obvious, but uncovering the answers took me years, and all the while most of my colleagues in the field did not agree with me. I failed to convince them, and in fact most physicians still see gastritis as a pathological condition. To them, a normal stomach should never show inflammation. The crux of the dilemma is simple: What is normal?

When pathologists see a stomach mucosa loaded with lymphocytes and macrophages, they call it chronic gastritis. But this condition can also be defined as the physiological response to our indigenous organisms. Just as there are inflammatory cells in your colon and in your mouth interacting with your friendly bacteria, your stomach has inflammatory cells interacting with its local bacteria. Thus the same question arises: Is the gastritis caused by *H. pylori* good for you or not? Pathologists, who characterize gastritis as a disease, classify *H. pylori* as a pathogen, whereas ecologists look at ancient organisms in an entirely different light.

The interaction between *H. pylori* and our ancestors evolved in ways that maintained persistence of the organism. Since there is little or no cost during childhood or young adulthood for carrying *H. pylori*, there is no selection against it. In contrast, malaria is so lethal to children that over the eons a whole set of human genes evolved to enable us to resist it.

We and our ancient, more quiet microbes like *H. pylori* are always adapting to each other, maintaining an equilibrium like artists on a tightrope, holding arms out as they cross to the other side; no missteps and they're safely across. Our microbes take up residence in particular niches and send signals to our human cells, which signal the microbes back in the form of pressure, temperature, and chemical messages, including defense molecules. The microbes signal us, we signal them back—communication develops, a language. Within this equilibrium, there is a dynamic of up and down regulation of inflammation in specific locales. It's like a marriage: we decide who does the dishes, who walks the dog. The conduct of one partner determines that of the other.

For example, the amount of inflammatory traffic in the stomach determines immune responses. Maybe the interactions early in life, when a baby is developing, also help determine immune tone.

A person's immunity can be twitchy, causing wheezing in response to an insect crawling across his or her arm, or it can be sluggish, with little response to a pathogen. There is no one universal tone in which one size fits all. Yet we have evolved over the millennia to have particular tones; it is not random. With our changing gastric microbiome, twitchier interactions seem to be increasing.

The loss of *H. pylori* from a person's stomach has created a new milieu. Instead of the ancient equilibrium, now the regulation of immunity, hormones, and gastric acidity is a dance without a partner. And like the ending of most long-term relationships, the effects are not just immediate or local; they are lifelong.

The changes that have happened in this past century have bearing beyond the stomach. At the very least, they affect the nearby esophagus, the next stop in the saga. New diseases related to the loss of *H. pylori* are rising.

10.

HEARTBURN

More than 60 million Americans experience heartburn at least once a month, and another 15 million have symptoms every day. If you're among them, you're in good company. Remember Bill Clinton's hoarse voice when he was in the White House? He suffered from acid reflux, the medical term for heartburn. So did George W. Bush when he drank coffee or ate peppermints. Star quarterbacks Brett Favre and John Elway played football with heartburn as did baseball greats Jim Palmer and Nick Markakis. When singers choke up and have trouble performing, the problem often can be traced to discomfort in their esophagus. So what exactly is this part of human anatomy and what causes it to give so many people grief?

The esophagus is a tube about eight inches long that connects your throat to your stomach. Just like the stomach, its entire length is lined with slippery mucus that helps food slide down. Every time you finish chewing, a bundle of muscles near the top of your esophagus open and you swallow what was in your mouth. If you take a swallow now, you will feel them.

Another group of muscles sit at the bottom of your esophagus and provide an opening to your stomach. When food collects in your esophagus, this sphincter opens so that your meal can drop into your stomach. When your esophagus is empty, it closes. In this way, food progresses in an orderly fashion down into your stomach, essentially as a one-way street. While swallowing is under your conscious control, the opening and closing of this lower sphincter is not.

When your esophagus is in good working order and you are not in the act of eating, this lower sphincter stays shut; stomach acid and stomach contents can't back up into your esophagus. But if it fails to close completely, you experience reflux. Acid creeps up the tube, leading to a burning sensation.

Reflux often comes and goes and is not, by itself, such a terrible problem. Wait for it to pass or take a couple of antacid tablets, and you're good to go. But when reflux becomes chronic, you're in danger of developing GERD, or gastroesophageal reflux disease, which can be extremely unpleasant and cause reflux every day. In addition to heartburn, patients may experience nausea, regurgitation, difficulty swallowing, and chest pain. The esophagus is irritated and ultimately may become scarred. GERD is now one of the fastest-growing health complaints in the developed world, affecting between 10 and 20 percent of adults in the United States.

One of my first hints that *H. pylori* might play a role in esophageal disease came out of left field. Recall that in 1987 Guillermo Pérez-Pérez and I developed a blood test for *H. pylori* and that I had tested positive, although I had no symptoms. And recall that a few years later we used my serum to identify a protein made by a particularly virulent group of *H. pylori* strains that are found most often in people afflicted with ulcers. By 1993 we found that this protein, CagA, is also involved in stomach cancer.

My father had an ulcer. My mother was from eastern Europe, where stomach cancer rates are high. Was I in danger of getting stomach cancer because of my family history? I felt fine, even though

I had the *H. pylori* strain most closely associated with ulcers and stomach cancer. But obviously, if I had believed the results of my own studies, I should have taken antibiotics to eradicate *H. pylori* and see what happened. Why take a chance on getting a horrible disease if I could prevent it?

It was time for a team effort. I asked my colleague Richard Peek, who had just finished his fellowship in gastroenterology, to perform an endoscopy. He would run a tube through my nose, down through my throat, through my esophagus, into my stomach. He would look closely as he passed it down and as he withdrew it. Through the tube, he would pass a tiny scissorlike apparatus, so he could snip biopsies of my stomach.

Another colleague, John Atherton, who came from England to work with us on *H. pylori*, would process the biopsies and isolate my particular *H. pylori* strain on a culture plate. And Guillermo Pérez-Pérez would again examine my blood for its antibody levels. Then I would take a course of antibiotics directed against *H. pylori*, and we would see whether over time my antibody levels would decline.

I hadn't reckoned on the discomfort the procedure would cause. On the day of the biopsy Rick gave me a drug to relax and reduce my ability to remember much of anything, and it worked, except for one thing: each of the seventeen times he passed the endoscope into my stomach to obtain a biopsy, I gagged. Why so many biopsies? We were researchers, and since I was willing, why not get a lot of material for all kinds of future studies.

Once it was over, they told me that they hadn't seen an ulcer or anything abnormal. I expected as much, but it was still good news. John Atherton put three bacterial samples taken from my stomach onto plates and waited for them to grow. Then I took antibiotics for ten days.

We waited. And we waited some more.

To my surprise, nothing grew. No lush colonies of *H. pylori* appeared in the bacterial samples. Blood tests had indicated that I carried *H. pylori,* but where were they? We guessed that I had carried relatively low numbers of the organism despite my high antibody values on the blood test. Or maybe it was because the high levels were suppressing the organism but not eliminating it, a little like how an oyster coats a grain of sand to produce a pearl. The oyster can't eliminate the sand, but it can make it less irritating. Over the next year we took a new blood specimen almost every month, and Guillermo found that my levels of antibodies reacting to *H. pylori* went down progressively and substantially, just as would be expected after successful antibiotic therapy. So now I could breathe easy. My risk of getting stomach cancer fell to almost zero.

Then something strange happened. About six months after my *H. pylori* were eradicated, I started getting heartburn after meals or in the evening, something that I had never suffered from before. I began to wonder if the heartburn was connected to my taking antibiotics. At medical conferences, I had heard anecdotes from doctors about heartburn being an occasional side effect of an antibiotic regimen, but it had never been properly studied.

GERD, unfortunately, can cause much more serious problems if left untreated. It can lead to a form of tissue injury known as Barrett's esophagus, which can then progress to the form of cancer called adenocarcinoma. In the past, nearly all esophageal cancers involved malignant changes in the upper and middle esophagus, closer to the mouth, and of a type different from adenocarcinoma. But ever since it was first identified in 1950, Barrett's disease has been found to sometimes progress to adenocarcinomas, involving the lower esophagus or the upper part of the stomach. Once a rare disease representing only 5 percent of all esophageal cancers in the United States, esophageal adenocarcinoma now has the fastest-rising incidence of all major cancers, a sixfold increase in the past three decades. It currently comprises more than 80 percent of all new esophageal cancer cases in this country and is also increasing throughout the developed world.

We didn't know these statistics at the time.

Despite many theories, no one had the answer to why all of these related disorders were rising dramatically: GERD, the mildest and most common, starting in the 1930s; Barrett's, more advanced but less common, in the 1950s; and the much-feared adenocarcinoma in the 1970s, but they were clearly connected.

At that point, we were mostly studying how *H. pylori* were injuring the stomach. Rick Peek, the doctor who endoscoped my stomach, was investigating how CagA-positive (the most virulent) and CagA-negative (less virulent) strains differed in their effects on the stomach. Since associations had been made between *H. pylori* and so many different diseases, I asked Rick to look at their relationship with GERD. We could use our blood test to address whether GERD patients had *H. pylori* more often than people with a normal esophagus. Working with colleagues at the Cleveland Clinic who specialized in GERD, Rick rounded up a collection of serum specimens, and Guillermo ran the blood tests, blindly as usual, not knowing which specimens were from normal persons and which were from those with GERD.

Surprisingly, instead of finding a positive association between *H. pylori* and reflux, Rick saw an inverse relationship. The patients *without H. pylori* were about twice as likely to have GERD. Later studies found a ratio of eight times more likely. What could explain this?

I asked Rick about the association with CagA in the patients we were sampling, since we knew then that CagA-positive strains were more virulent. He told me that the association with Cag was even stronger and also in the inverse direction: less Cag, more GERD. This was the exact opposite of what we expected.

I actually knew very little about GERD at the time, and so I asked Rick whether GERD was increasing. When he confirmed that it was indeed, our *H. pylori* work moved in a new direction.

That first study formed an early basis for the hypothesis that *H. pylori* may be protective against GERD. The inverse association was there, of that we were sure. But what did it mean? What caused it to happen? How could a microbe living in the stomach that was involved in ulcers and cancer protect the esophagus? Or was it the disease in the esophagus that eliminated the organism?

For many years, a group in Germany had been treating patients who had duodenal ulcers with antibiotics to eradicate *H. pylori*. They began to study the consequences. Three years after the therapy, they examined each person's stomach and esophagus. In about half of the patients, the treatment worked—they no longer carried *H. pylori*. But in the other half, the treatment failed, and *H. pylori* persisted. This was a common outcome in the early days after antibiotic therapy was developed to eradicate *H. pylori*. Today, doctors prescribe different regimens that generally have higher success rates—above 80 percent.

But in comparing the two groups, the German scientists found that of those in whom *H. pylori* remained 12.9 percent had reflux, but almost 26 percent of those with successful treatment, who now were without *H. pylori*, also had the problem. The elimination of *H. pylori* led to double the rate of esophageal disease. This was pretty remarkable and provided evidence of the direction of the causal relationship: therapy that eliminated *H. pylori* worsened the esophagus by promoting reflux.

Many other researchers in the area attacked this paper on a number of technical grounds, and over the next year or so it was fashionable at conferences to denounce it. But it caught my attention. I knew that the lead investigator, Joachim LaBenz, was a serious scientist of great integrity.

Over the next few years, my group carried out additional studies with colleagues from around the world and found the same dynamic: inverse associations between *H. pylori* and GERD, Barrett's esophagus, and adenocarcinoma. People who had the more virulent *cagA*-positive *H. pylori* strains, the ones associated with ulcers and stomach cancer, had the highest degree of protection against diseases of the esophagus.

This presented quite a puzzle. How could *H. pylori*, the bad guy, protect the esophagus, and how could the *cagA*-positive strains, the most virulent ones, be especially protective?

We can look to stomach acid for clues. Acidity kills most bacteria. But over the eons of evolution, *H. pylori* has found ways to avoid annihilation in an acid environment. In a sense, *H. pylori* actually likes acidity, because even though there are costs associated with living in such a hostile environment, acid keeps away competitors. The enemy of my enemy is my friend.

In fact, a body of research from many labs, including mine, was showing that *H. pylori* help regulate stomach acidity by causing inflammation that affects stomach hormones that in turn switch acid production on and off. Over the first decades of life, this acid-balancing system works pretty well. Under the microscope, the glands that make acid resemble fronds waving in the breeze. But as a person gets older, chronic inflammation starts to wear down the stomach walls, and in those who have *H. pylori* they wear down faster. The glands that make acid begin to shorten and flatten out. When this occurs the stomach develops what we call atrophic gastritis—it makes less and less acid. As a consequence, ulcers tend to go away. Schwarz's dictum—no acid, no ulcer—remains correct.

But people who never acquired *H. pylori* in childhood or who had their bacteria knocked out by antibiotics have high levels of acid well past the fortieth year of life. Thus, for possibly the first time in human history and prehistory, large numbers of people are reaching middle age with fully intact acid secretion. For them, the stomach contents moving up into the esophagus are highly acidic and have more digestive enzymes—and are more damaging. And with the markedly decreased prevalence of *H. pylori* in childhood, most of today's children are growing up with different acid regulation than the children of prior generations had, with *H. pylori* out of the physiologic picture. Reflux in children, once ultra-rare, keeps increasing, and many children today are being treated with medications to reduce gastric acid levels. Could these events be connected?

We were finding that *H. pylori*, discovered as a pathogen, is really a double-edged sword: as you age, it increases your risk for ulcers and then later for stomach cancer, but it is good for the esophagus, protecting you against GERD and its consequences, including a different cancer. As *H. pylori* is disappearing, stomach cancer is falling, but esophageal adenocarcinoma is rising. It is a classic case of amphibiosis. The facts are consistent.

11.

TROUBLE BREATHING

Most people are aware that asthma, a disease recognized since antiquity, has become a massive health problem. Statistics from developed countries, where records have been kept for the past seventy years or more, show that rates have been doubling and tripling. The graphs look like what you would want for your pension fund but instead indicate an increase in terrible suffering and sometimes early death.

Physicians have known for many years that GERD and asthma are somehow linked. Many GERD patients develop the wheezing, coughing, and constricted airways emblematic of asthmatic attacks. And when asthma patients receive treatment for GERD to reduce stomach acidity, their breathing often improves. Despite the connection, most physicians believe that GERD accounts for only a small fraction of asthma cases.

One theory that explains how the two disorders are related is purely mechanical. When stomach acid travels up the esophagus, it can spill down the windpipe, creating irritation. But that explanation fails to account for the allergies and hay fever often associated with asthma. Asthma is the leading edge of a group of related disorders that involve too much sensitivity to foreign substances.

After our studies showed that *H. pylori* can protect us from GERD, I began to wonder whether it protects us from asthma as well. Maybe the rising incidence of asthma is related to fewer children acquiring *H. pylori* early in life and more children having it eliminated inadvertently with antibiotics. Could subclinical unrecognized cases of GERD caused by the lack of *H. pylori* be driving the asthma epidemic?

Although it made sense and was consistent with what we were learning at the time, the mid-1990s, the connection presented a major and controversial leap. The fall of *H. pylori* and the rise of asthma could both be true but unrelated, just as the rise of asthma parallels the rise in home TV sets or the number of Volkswagens on the road.

I attempted to get several colleagues who worked in lung diseases interested in studying this potential relationship with me, but it was too far-fetched, and besides the medical community at large was focused on *H. pylori*'s dangers. To investigate my hypothesis, I needed to study a population of patients with asthma, but without a collaboration with a clinical scientist working in that area, it would be impossible.

Then, in 2000, I moved from Vanderbilt in Tennessee to New York to become the chair of medicine at New York University. It was a great opportunity to return to my alma mater and to help build a strong department. But despite the administrative tasks and pressure, I didn't want to give up research. At a new place, I had new chances. I asked my colleagues, "Who works on asthma here?"

Everyone pointed to Dr. Joan Reibman, a specialist in lung diseases, who in 1991 had established a clinic at Bellevue Hospital for adult asthma. Joan listened politely to my ideas, but with little enthusiasm. One of Joan's great intellectual strengths is her skepticism. Wild ideas come up all the time. She wasn't buying mine unless I could show her evidence. Fair enough.

Joan agreed to use the clinic she started at Bellevue Hospital to enroll patients in the type of study we wanted. Their friends and relatives who did not have asthma would serve as the controls. She performed a battery of tests to characterize their lung function and their allergies. Fortunately for me, beginning in 2002, she also collected and froze blood specimens from the participants that we could use to assess their *H. pylori* status. Joan's support was critical to test the hypothesis; she was and still is dedicated to finding answers for how to improve care for patients with asthma.

By 2004, Joan's team had collected blood from more than five hundred people. We agreed that she would send my team the serum specimens under code to blind us from knowing who had asthma and who was a control. This removed chances of unintentional bias in the analysis. Guillermo ran the blood tests, and then we divided the results into positive, negative, and uncertain, and through repetition we resolved essentially all the ambiguities. Later that year, we sent the results to Joan and her team, which included Michael Marmor, a seasoned epidemiologist accustomed to the type of statistical analyses we needed. Joan called a few weeks later and told me that, much to her surprise, she and Mike had found an inverse association between *H. pylori* and asthma. Still she was dubious. After all, how could a stomach microbe protect against asthma?

We agreed to meet and review the results. A week later, Joan, Mike, and the rest of their team came to the VA hospital where my lab is and where I have a small office. Joan described the participants—318 patients with asthma and 208 healthy controls—and announced that the statistical analysis showed that people positive for *H. pylori* were 30 percent less likely to have asthma than people without the organism. This was true even when they took into account variables that could otherwise explain the propensity for asthma.

This was the first, early support for my theory. Even so, there were many possible ways to interpret the data.

"What about *cagA*?" I asked. We had run blood tests for both *H. pylori* status, to see who was carrying the organisms, and also for *cagA-*

positive strains, just as we had done for ulcer disease, stomach cancer, and esophageal diseases.

"We haven't analyzed that yet," Joan said.

I was disappointed because *cagA* is really the key marker. The *cagA*-positive strains are the ones that are the worst to have in relation to ulcers and the best to have in relation to a healthy esophagus. If I had to predict where the strongest story for asthma would be, I would be betting on *cagA* as the best predictor for protection against asthma.

"Well," Joan said. "We'll just have to look at that later."

Then Mike interrupted. "Wait a minute!" he said. "I should be able to get that."

With that, he began to type on his laptop. We all watched in silence. After about thirty seconds of typing, with a flourish Mike pushed a final button. And a few seconds later he read from the screen: "*cagA*+; odds ratio of 0.6."

Eureka. That meant that people carrying *cagA*-positive strains of *H. pylori* were 40 percent less likely to have asthma than people without *H. pylori*. This was stunning.

The strains most related to stomach cancer and ulcers turned out to be the most beneficial for GERD and now for asthma. It seems like a paradox, but the finding now can be explained by the fact that *cagA*-positive strains are the most highly interactive with their hosts. By then, we understood how these strains operate: they are constantly injecting their own materials into human stomach cells. It is as if there are two different populations of *H. pylori* strains. Some are vigorous and highly interactive—the *cagA*-positive strains. The others, the *cagA*-negative strains, may be regarded as more sluggish; they have much less contact with the cells of their human host.

The *cagA*-positive strains probably are living a little closer to our cells, while the others are further offshore in the lumen. Thus, not surprisingly, the *cagA*-positive strains are the most damaging to the stomach wall. But because they are the most interactive, they *also* have the potential to be the most beneficial in helping regulate our physiology.

Next, Joan examined the clinical records of the asthma patients to learn how old they were when their disease was diagnosed. Were they children or adults when their symptoms first appeared? We found that H. pylori–positive subjects were, on average, twenty-one years old when their asthma started. For those without H. pylori, the average age of onset was eleven. This was a striking difference. It showed that lack of H. pylori was more commonly associated with childhood-onset asthma and suggested that in people who were bound to get asthma, the presence of H. pylori might delay the process. A couple of years later, a large study was done among children in Manitoba, Canada. The investigators found that antibiotic use in the first year of life was associated with a significantly greater chance of having asthma at the age of seven. They weren't looking for H. pylori, but their findings were consistent with my general hypothesis.

In Joan's study, the blood samples also were tested for antibodies to allergens, so that we could examine whether having H. pylori correlated with allergic responses. Here again we saw a connection: the presence of H. pylori was associated with fewer reactions to the allergens, suggesting that H. pylori can protect against allergy.

An abstract of our findings was submitted to the 2005 annual meeting of the American Thoracic Society and presented in May of that year. Unfortunately, it was received with a huge yawn. Our work was outside the mainstream of asthma studies, and even the pulmonary disease specialists on Joan's team were not as impressed by the results as I was.

■ ■ ■

I pressed on. Could we reproduce our findings in another population? If the findings were real, we should see them again. I thought about using a large study, called NHANES III, in which the United States selected twenty thousand people as a representation of the overall population and between 1988 and 1994 gave them a series of health exams.

The results of blood tests were still available, including the subject's *H. pylori* status. Sitting in the same little office at the VA where I had met Joan's team some months earlier, in March 2006 I suggested to Dr. Yu Chen, a young epidemiologist new to NYU, that we use these data to test the hypothesis of an inverse association of *H. pylori* and asthma. Yu agreed and was able to find records from NHANES III with information on the asthma status and the *H. pylori* status of more than 7,600 people. Joan's study, involving 500 people, was large as studies go, but this one was about fifteen times bigger.

On May 5, 2006, Yu e-mailed me: "I have done some analysis using the NHANES data . . ." she wrote. ". . . It's kind of strange."

Rushing to catch a plane to Chicago, I stuffed the tables she sent me into my briefcase. A few hours later in the quiet of the cabin I took them out. The results were clear: Yu had done a terrific job, and her analyses showed that there was an inverse association of *H. pylori* and asthma in NHANES III. It was especially pronounced for *cagA*-positive strains. In fact, the percentage margin was just about 40 percent, almost identical to that of Joan's study.

Here was a second large, independent, blinded study that had shown almost exactly the same result as the first. This could not be explained by random chance. Although other issues needed to be considered, and these data did not indicate whether lack of *H. pylori* predisposed to asthma or vice versa, sitting in the plane with the hum of the engines outside and my neighbor asleep next to me I knew at that moment that my hypothesis was correct. I felt I had hiked up a very long trail and finally, gasping for breath and soaked in sweat, I reached the summit of the peak I had been climbing—it was a moment of exhilaration.

Yu's study showed more nuance. All of the inverse association was observed in children under the age of fifteen. The effect was specific for childhood-onset asthma but there was no effect for adult-onset asthma. Although incidence of asthma has increased since World War II, it has risen most dramatically among children. Childhood-onset asthma affects kids in cities and in the country, all over the developed world, but the poor are especially susceptible. There are many theories to explain this, but a popular one is that poor children are more heavily exposed to cockroaches and other insects, which may be important triggers of asthma. But not every child in a cockroach-infested home becomes asthmatic, and plenty of children get asthma with nary a cockroach in sight. To me, the mystery is not why someone becomes allergic and wheezes after being exposed to a cockroach. That part I understand. Rather the question is why after the exposure, when most children stop wheezing, other children do not.

The NHANES III records also contained data on hay fever, or allergic rhinitis. Again we found an inverse effect, again in children but not in adults, again stronger for *cagA*-positive strains. This work provided the first indication that the presence of *H. pylori* in a child's stomach might protect against hay fever. As with asthma, hay fever has been getting more common in children, as *H. pylori* has been disappearing.

NHANES III was a treasure chest (your tax dollars actually at work). Yu was able to link *H. pylori* status with results from allergy skin tests on more than twenty-four hundred people. For each of the six allergens studied, there was an inverse association of sensitivity with *H. pylori*, and for four of them (ragweed, rye, thistle, and alternaria) the differences were statistically significant. As with asthma and hay fever, people with *H. pylori* were less likely to have skin reactions to allergens. To be clear, I'm not suggesting that there is any direct relationship between *H. pylori* and, let's say, thistle. But rather it appears that *H. pylori* is having some general effect on immunity, on people's ability to turn off an allergic response.

These additional findings were very important, showing similar relationships between each of three different but related diseases— asthma, hay fever, and skin allergies—and *H. pylori*. They also confirmed our findings in larger samples. This happened again when Yu and I performed another large study that involved samples from people enrolled in NHANES in 1999 and conducted about a decade later, which showed highly consistent results. In scientific research, a single study is rarely enough to prove a point. It's always better if multiple studies show consistent results.

I had begun to study asthma because of its relationship with GERD. I had gone along with the popular explanation that reflux could lead to asthma by exposing the lower esophagus to acid, bile, and other toxic substances, which then might move up the esophagus and then down the trachea into the airways. But that theory does not explain the incidence of hay fever and skin allergies, since their primary sites are far away from the esophagus. Considering that these maladies are all allergic disorders, the obvious question became whether *H. pylori* influences immunity. And then: How can a stomach microbe affect a person's immune status?

The answer I eventually came to can be traced back to the original observations of Robin Warren, the Australian pathologist who linked *H. pylori* to gastritis. Gastritis is a large accumulation (more than what is considered normal) of inflammatory and immune cells in the stomach wall. But which stomach wall is normal, the modern wall with no *H. pylori* and small numbers of such cells (no gastritis) or the more ancient wall with *H. pylori* and large numbers of those cells (gastritis)?

Like your intestines, your stomach wall is home to numerous cell types involved in immunity. These include white blood cells that fight infections and others that regulate immunity. Your stomach also has so-called dendritic cells with long protruding processes that sense and respond to nearby bacteria. When these dendritic cells are activated by bacteria or their products, they sound the alarm to lymphocytes, white blood cells that are a central part of your body's police force.

Lymphocytes ramp up defenses in many ways. They also have an intelligence function; they have memory. Each of us has an army of memory cells, most of which remember some chemical aspect of a particular event, such as a component of a bacterial wall from a prior infection. Memory allows your body to accelerate responses to danger, to bring out the reinforcements for something that is a remembered threat when it reappears. Every time children get a strep throat, their bodies gain more memory of the components of the bacteria, and eventually they stop having symptoms, even when reexposed. They have become immune. Vaccines and their booster doses take advantage of our memory functions to provide strong immunity.

It's not surprising that the wall of your gastrointestinal tract, from your mouth to your anus, is populated by dendritic cells that sense bacteria and by lymphocytes that react to bacteria. These defenders respond to your regular microbial residents and to unwelcome invaders but not always in the same way. These lymphocytes remember the usual suspects that need rounding up when they are detected or the ones to be treated with kid gloves.

Your stomach wall has its own populations of lymphocytes: B-cells that produce antibodies and T-cells that orchestrate the complete defenses. But like your arm muscles, which are comprised of biceps that flex and triceps that extend, immune cells can have opposing functions. They can be activators or suppressors. Some T-cells predominantly turn up inflammation, whereas others, called regulatory T-cells or T-regs, modify and suppress the responses. We don't want every minor incident to erupt into full-scale war; that would be too destructive. We need a police force to regulate the army, like military police, who keep order among the troops. That's one role performed by T-reg cells, which along with the activator types of T-cells live in the stomach's wall. Part of the "gastritis" that pathologists see in the stomach wall when people are colonized by *H. pylori* is the lymphocytes that are reacting to the organism. There are many more lymphocytes and far more T-reg cells in the *H. pylori*–positive stomach than in the modern *H. pylori*–free stomach. And they are there doing their job, calibrating inflammation.

As such, the gastritis observed by pathologists is not all bad.

This is a paradigm shift. I believe that the T-reg cells that reside in your stomach help, through their suppressive functions, protect you from asthma and allergic disorders. Pathologists and physicians need to recognize that "inflammation" of the stomach is normal. It has a biological cost in terms of ulcers and stomach cancer, but it has biological benefits that we are just now recognizing.

A Swiss group led by Dr. Anne Mueller has conducted important experiments in mice to understand the significance of *H. pylori*–induced immune responses. Their work strongly supports a protective role of *H. pylori* in asthma. Mueller and her colleagues induced asthma in mice by spraying an aerosolized allergen into their lungs. They showed that when the mice were infected with *H. pylori*, their responses to the allergen were reduced. Having living *H. pylori* in their stomachs protected them against the asthma; giving the mice dead *H. pylori* cells conferred no benefit. Moreover, the mice infected with *H. pylori* at a very young age gained greater protection than did the mice infected as adults. These findings parallel the epidemiology in humans. We showed that most *H. pylori*–associated protection against asthma occurs early in life.

Further experiments in mice by Mueller's group showed that *H. pylori* interact with the sensing dendritic cells of the stomach wall, causing them to program the immune system to turn out T-reg cells. What a smart strategy for *H. pylori*; the T-reg cells suppress immune responses that act to eliminate it. But it is a grand bargain because a collateral benefit to us is the suppression of allergic responses, as Mueller and her colleagues showed in mice.

This theory, while not yet well known, makes evolutionary and physiologic sense, and the epidemiologic, histologic, and experimental studies provide parallel lines of evidence: *H. pylori*–induced immune-cell populations protect against asthma. Again, the idea is not that *H. pylori* have anything to do with cockroaches or with ragweed. Rather, their presence early in life helps ensure that when the host encounters them, it should be able to turn off its immune responses before the allergies get out of hand. And this dynamic may not be limited to just *H. pylori*. There might be other microbes that have disappeared and with them their dependent immune-cell populations. *H. pylori* may be the bellwether, the lead sheep in the flock, the main actor in a large company, or it is the star of the play, or a one-man show. We don't yet know. But these old troupers are fast disappearing, which may be sufficient to explain the growth of asthma.

■ ■ ■

My ideas about *H. pylori*—that they appear to be beneficial early in life for health and well-being but dangerous to health in later life— have not been well received by many of my colleagues. Just the opposite. Some have even labeled me a heretic.

A big part of the problem is that *H. pylori* was discovered as a pathogen, and a significant edifice has been built about it being bad for you. In part, this resistance appropriately reflects adherence to a beloved scientific principle: showing an association is not the same as showing causation. People who rob banks may smoke more than those who live ordinary lives, but that does not mean that smoking causes them to rob banks. In fact, there may be "reverse causation." Robbing banks is stressful, and people might smoke more to calm their nerves.

Despite the many studies from multiple investigators examining different populations, direct proof for *H. pylori*'s dual nature is limited. Still, the degree of skepticism is out of proportion to the nature of the evidence. For example, a causal role for *H. pylori* in ulcer disease has never been shown. Investigators have shown that eliminating *H. pylori* markedly reduces the risk of ulcers recurring, and that is extremely important clinically but doesn't address what caused the ulcers originally.

Imagine that I spilled gasoline on my hands and that someone lit a match. My hands would wind up with burns. Let's say that we conducted a study of treatment and applied an antibiotic salve to my right hand but not my left and that the right hand healed better. What could we conclude? Clearly, applying the antibiotic improved the outcome. If we did this trial with many people, and on average their treated hand responded better than the untreated one, it would become the new standard for care.

But such a trial would not prove that bacteria caused the burn. Only that their removal improved recovery. The burn was caused by the conjunction of the gasoline and the match. After-the-fact studies of treating *H. pylori* in people who already had ulcers are just like that. Actually, the only study I am aware of that examined whether *H. pylori* might precede ulcers was ours. Again working with Abraham Nomura and the population of Japanese-American men in Hawaii, we showed that having *H. pylori* in the 1960s was associated with an increased risk of developing an ulcer in the next twenty-one years. So I am not at all against the idea that carrying *H. pylori* has important costs to us. It's just that like many complex problems in human biology, the specifics of causation are hard to pin down. And while *H. pylori* is generally necessary for ulcer disease, it is far from sufficient. In 1998, I proposed that ulcers are due to the changing microecology of the stomach, meaning changes in the numbers of *H. pylori*, in the types of strains, in the variety of strains, and in the other organisms present and their distribution. Sixteen years later, the idea still looks valid.

Following the initial work by Warren and Marshall, a group of "helicobacteriologists" emerged. Meetings were held all over the world, and a group of us acquired many stamps in our passports. Each year the European community holds an annual *H. pylori* workshop, with attendance by gastroenterologists, microbiologists, pathologists, and their students, who grew in number to the thousands by the mid-1990s. Ample support from pharmaceutical companies eager to identify with the new "movement" provided important fuel and rationale for these meetings.

In 1996 and 1997, when I presented the idea that there might be "good" helicobacters, reactions ranged from amused tolerance to disdain. Remember, the only good *H. pylori* are dead ones. Warren and Marshall's Nobel Prize in 2005 didn't help my cause either, although the Nobel Committee was very precise in its citation, which referred to both the discovery of *H. pylori* and to its role in peptic ulcer disease. The great revolution created by the discovery of *H. pylori* was to overturn the dogma that ulcers were due to stress and its consequent hyperacidity. Now a new dogma, that *H. pylori* should be *eradicated*, became as firmly entrenched.

Among doctors who felt they were doing an obvious good by eliminating *H. pylori*, patients who were worried about the "infection," and the pharmaceutical industry, which was happy to sell its products, including acid-suppressive therapies (among the best-selling drugs in the world), a steamroller had formed to flatten the ancient microbe. Even though relatively few people actually had ulcers at any one time, the momentum continued to build.

Nonetheless, I believe that ultimately we will understand that an ecological shift of this magnitude—the disappearance of *H. pylori*— must have many consequences, good and bad. All along, my work with *H. pylori* has shaped my thoughts and has led me to where I am today, concerned about the disappearance of many microbes from our ancestral bacterial heritage. How many other organisms have disappeared or are disappearing?

My colleagues continue to have "consensus conferences," largely underwritten by pharmaceutical companies, that keep extending the umbrella of who needs to have *H. pylori* eliminated. "Test and treat" is still the practice of the day. The military analogy would be "search and destroy." People everywhere are scared of the *H. pylori* in their stomachs, and doctors feel they are duty bound to remove this pathogen. Despite publishing our results in many leading journals, I have not been able to budge the clinical dial.

However, my ideas have resonated in the broader community of microbiologists and ecologists. Because of the role my team has played in putting *H. pylori* on the map as a pathogen, I am invited to many meetings and universities and have been inducted into leading scholarly societies. In my articles, I stopped calling *H. pylori* an infection. I call it a colonization, just like the colonies of countless other organisms that live in your body, mostly happily, for years. About that, I am certain.

I am also convinced that time is on my side, that truth will out, and that we will learn to give more personalized treatments—to decide who should have *H. pylori* eliminated, who should keep it, and who should have it restored. We are moving in the right direction, but today's medical practice has many counterproductive incentives and a lot of inertia, especially when sacred cows are involved.

12.

TALLER

We were on a back road, taking what on the map looked like a very direct route to Chichén Itzá, a large pre-Columbian city built by the Mayas. It was dry and dusty, but the dirt road was good. Now and again, we could see the roofs of houses through the bush. Other than the road, there wasn't much evidence of progress amid the seared landscape. Yet the Yucatán had once been one of the centers of a civilization encompassing millions of people, stable over centuries. Now, not far from the ruins of a great ceremonial site, the land was mostly scrub forest, desolate and monotonous.

But down the road I could see two children. As we got closer and drove past them, their faces came into view. They were purely Mayan, with jet-black straight hair and broad, smooth features, the type we see in murals or sculptures on a classical Mayan stele. But something was immediately askew. These kids, maybe eight and eleven years old, were much too heavy. They were obese. I could expect to see obese children on the roads of Arkansas, Ohio, or Bavaria, but here in the Yucatán it was a shock.

"It's even happening here," I said to Gloria, who was traveling with me. She knew that I was studying obesity, so my implication was evident. I was surprised how far the epidemic had spread, reaching even to remote areas in developing countries. Later, when I related what I saw on the road to one of my colleagues at NYU, he told me that he had observed the same thing in Ghana: "When I started working there more than thirty years ago, the major problem in children was malnutrition. Now it's obesity."

Why are people all over the world getting fatter? For the first time in human history, overfed people outnumber the underfed. Globally, one in three adults is overweight. One in ten is obese. By 2015, the World Health Organization estimates that the number of chubby adults will grow to 2.3 billion, equal to the combined populations of China, Europe, and the United States. Children and adolescents are also heavier everywhere we look. Are people eating too much junk food and exercising too little?

As a doctor and scientist who studies human health, I am both disturbed and fascinated by this question of why people are getting fatter. And I have found what I believe are some promising leads for answering it. But before I get to them, I want to discuss a related question that led me, circuitously, to those answers: Why are people all over the world getting taller?

Average human height has been increasing in many countries for the last one hundred years. When I ask most people why they think this is happening, they say it is because of better nutrition, and it is hard to argue with that. In developed countries, we are certainly eating more than our ancestors did, although whether our diets are better is a whole other issue. Famine is mostly a thing of the past. In that sense, nutrition has clearly improved, and I am not diminishing its importance. But, as always, I have been most interested in the contribution of microbes to how humans develop.

Some years before, I remembered a study carried out between 1964 and 1973 by Leonardo Mata, a microbiologist and public-health expert at the University of Costa Rica, on the relationship between malnutrition and infection in the children of Santa Maria Cauque, a rural community in Guatemala. Back then childhood mortality was staggeringly high in Guatemala, about 96 deaths per 1,000 births compared with our 6 per 1,000 in the United States today. Sanitation was poor and the children suffered from a litany of diarrheal illnesses. Mata and his colleagues found that the more often the kids had diarrhea, the more slowly they grew. With more disease, they were shorter. Mata's work was consistent with a wide body of data, but his study in particular caught my attention because his findings were so clear.

It's widely believed that the period of maximal growth velocity (otherwise known as the growth spurt) occurs during adolescence, but that is not the case. The first two and half years of life, the period with the greatest velocity, is the main critical window for the development of adult height. Experienced pediatricians know that if you double a child's height at age two, you can closely estimate how tall the child will be. Studies of children adopted from Asia showed that if they moved to America before their third birthday, they grew to the average height of their new playmates. But if they moved later, then they retained their height status from the old country. Thus, the place to look for factors that affect height is early childhood.

Another key observation on what influences height came from studies of *H. pylori*. Soon after the microbes were discovered, scientists began searching for associations between them and all aspects of human health. For example, people who carried *H. pylori* in their stomachs were more likely to have been impoverished in childhood. Moreover, adults who had *H. pylori* were, on average, shorter than those without. The studies focused on the idea that *H. pylori* stunted growth, which was consistent with the view in those days that *H. pylori* was always bad for you. This research suggested that pathogens made people short. If you got rid of the pathogens, they would grow taller. It made sense to me.

By the 1990s we knew that *H. pylori* was acquired in the first few years of life, when it could plausibly make a difference in height. And *H. pylori* was associated with poverty in childhood, which also fit, since poorer people tend to be shorter. But no one knew whether *H. pylori* alone was stunting height or if it was a marker for other microbes, perhaps acquired by the same fecal-oral route.

We subsequently learned that *H. pylori* affects the regulation of the hormones ghrelin and leptin, both produced in the stomach and both involved in the storage and use of energy. We can imagine that young children growing up with *H. pylori* in their stomachs might be metabolically different from those without and that this hormonal variation could slow down their growth trajectory, making them shorter. This is a hypothesis that requires experimental support, but some of our more recent experiments with mice, described below, provide additional evidence.

In 2000, when I returned to NYU, I looked for someone to help me research the question of why people are taller. Albertine Beard, a medical student, took up the challenge and soon uncovered lots of interesting data. It turns out that long-term changes in height are relatively easy to measure; anthropologists use skeletons to estimate height, and armies have been measuring the height of their soldiers for centuries and keeping records of it.

Albertine found that the history and prehistory of humans has not been a long, inexorable path to greater height as we all might think. Skeletal remains indicate that people grew taller at various times in prehistory and history, and then got shorter. This pattern varied by locality and across time periods. We learned from U.S. Army records that the soldiers in George Washington's army in the eighteenth century were taller than the soldiers who fought in the Civil War in the 1860s. Why would that earlier generation have been taller?

More recent is the remarkable trend of increased height in the late twentieth century. The Dutch, who were among the shortest people in Europe in the early twentieth century, are now among the tallest. The streets of Amsterdam are filled with young giants, male and female. In Asia, the trend has been even more dramatic. When I studied in Tokyo in 1975, I saw a sea of black-haired heads when, at six foot two inches, I rode in the overcrowded subway cars. As I returned over the years, a face would pop up from the crowd on occasion and then more faces. Now, nearly forty years later, there are many tall young Japanese people, and to make things even more strange, their hair, thanks to chemical dyes and fashion, is blond, red, purple, and blue. In China, which had an increase in growth later than Japan, the average six-year-old boy in 2005 was more than two inches (6.5 cm) taller than his counterpart in 1975; girls had a similar (6.2 cm) increase. These are monumentally rapid changes.

These trends have many possible explanations, including better nutrition, but we developed a theory about how microbes might affect height. It is not that we think nutrition is unimportant, but it's not sufficient to explain the temporal and geographic patterns seen. As discussed in prior chapters, the nineteenth century was a time when sanitation first got much worse in industrializing countries, and then as a result of public-health measures it began to get much better. During the early nineteenth century, municipal water supplies usually contained a microbial soup of human pathogens as well as friendly or commensal bacteria, both of which are ingredients of fecal contamination. From the late nineteenth century onward, when water supplies were filtered and chlorinated in many parts of the world, pathogens began to be eliminated and people got healthier—and taller. There was less cholera and milder diarrheal illnesses. Vaccines controlled diphtheria, whooping cough, and other important infections of young children.

But it also is possible that observed changes in height may be due to the loss of friendly bacteria as well as these pathogens. Our understanding of the microbes inhabiting us is at an early stage, so we don't yet know the identity of microbes that could help make us taller or even if they exist, but based on our recent work I'm willing to wager that we'll find them.

This connection between microbial transmission and height does shed light on the question of why soldiers in the Revolutionary War were taller than those in the Civil War. If you were raised on a farm in the mid-eighteenth century, you would have grown up relatively isolated. Eighty years later, growing up in overcrowded urbanized America, you would have been subject to childhood epidemics, and your water was likely to have been more contaminated.

In 2002 we published these ideas and the supporting evidence in an article entitled "The ecology of height: the effect of microbial transmission on human height" in *Perspectives in Biology and Medicine,* a well-respected journal, but the article received very little attention—another big yawn.

Nevertheless, I was already thinking of the sequel, "The ecology of weight," with many parallel ideas in mind. As it turned out, I never wrote that paper because an alternative pathway to understanding why we are fatter became much more exciting. To begin that story we need to go back to 1979, when I went to work in the Enteric Disease Branch at the Centers for Disease Control, where I was the salmonella surveillance officer of the United States. My assignment was to track and study *Salmonella* as well as other bacterial pathogens affecting the GI tract. That's also the time I had my own serious *Salmonella* infection, as I described. Unlike my case from eating tainted watermelon, most of us get *Salmonella* infections from foods of animal origin, including meat, eggs, milk, and their products.

Recall that farm animals are given low (subtherapeutic) doses of antibiotics to promote their growth. At the time, no one was curious about why growth promotion was so effective. When writing the paper on height, I realized that a great experiment had been happening on our farms with results consistent with my ideas about the role of microbes in height and weight.

If farmers can purposefully enhance the growth of their livestock by giving antibiotics to young animals, what are we doing to our children when we give them many similar medications? Could our widespread use of antibiotics to treat infections in children be having analogous effects?

On the farm, the practice is deliberate, and animals are given low doses of antibiotics more or less continuously. Obviously it works; the animals gain weight. We give our kids much bigger therapeutic doses, but only episodically, to treat their infections. In one sense, this is a big difference, but in total, the idea is the same: exposure to antibiotics early in life causes microbial perturbations at a crucial time, just when organs and systems are developing. The idea that antibiotics might be causing weight gain in our children, that they could be a "missing link" in the obesity epidemic, seemed a reasonable possibility, but we would have to study it to see if it was true.

■ ■ ■

There is no question that antibiotic use fundamentally changes the development of the youngest animals. The earlier that farmers start to give antibiotics to chickens, cows, and pigs, the more they alter their development. Most significant is that farmers have found that virtually any antibiotic promotes the growth of their livestock. They all work, despite differences in chemical class and structure, mode of action, and the spectrum of activity for the microbes they target.

If despite their differences they all work, then it must be because of their effects on the microbiome in general, not because of their unique side effects or the specific bugs they target. The drugs must affect the very composition of the microbial community and the interactions they have with their hosts. They must be affecting many if not all aspects of growth and the development of metabolic systems during a critical window.

Also of great interest to me was that the earlier in life farmers started giving animals antibiotics, the stronger the effect. The simplest explanation is that antibiotics lead to a shift in the overall equilibrium of our gut microbes. Some bugs become more dominant; others are suppressed. As we know, microbes evolve along with each animal species they inhabit. Now farmers were deliberately changing the conditions under which microbes and their hosts coevolved to attain equilibria. As John Nash's model predicts, when equilibria are perturbed, bad things can happen. The idea was simple, but the implications were huge.

To explore how the practice of subtherapeutic dosing with antibiotics affects development, we began a series of laboratory experiments on mice. This has been the most exciting work of my career.

13.

. . . AND FATTER

Why do antibiotics make animals bigger and fatter? The goal in our study was to re-create the weight and size increases observed in farm animals in our lab and then tease out the principles for why they occur. It took a big team to address these questions, but several scientists played key roles: Ilseung Cho, a physician and a fellow in gastroenterology; Laurie Cox, a graduate student whose dissertation project revolved around the mouse models and who at age fourteen had started working with bacteria for her father's company, which made products for clinical bacteriology labs; and postcollege student Yael Nobel. Without such intelligent and dedicated trainees, I could not have tested any of my ideas. And there were many others who joined in the quest, from high school and college students working during the summer to college students doing independent research and visiting scholars from around the world.

In 2007, after a number of attempts to get the model going, we began our first complete set of experiments on farm practices by adding four different subtherapeutic antibiotic treatments, which we called

STAT, to the water bottles of mice. We only looked at females because they don't fight as much as males, making work easier for us. The early results were not promising; there was no weight difference between the STAT and control mice.

When Ilseung's research committee was told that the mice were not gaining weight, one of our experts asked, "What's happening to their body composition?" He was referring to the proportions of fat, muscle, and bone. We didn't know.

"Why don't you DEXA them and find out?" he asked.

DEXA them? The term refers to dual-energy X-ray absorptiometry, a test given to women to determine their bone mass and risk for osteoporosis. But DEXA also tells us how much fat is in the body and how much muscle.

This suggestion turned out to be critical. We discovered that all four groups of STAT mice had about 15 percent more fat than the controls, differences that could not be explained by chance alone.

We had our first evidence that antibiotics were changing metabolism, affecting body composition. The STAT mice were making more fat and had about the same amount of lean muscle as the control. We also had an unexpected finding: at seven weeks of age, three weeks following the start of the antibiotics, the mice were putting on bone at an accelerated pace. More bone formation implies that they would become bigger, longer, and taller. But by ten weeks, all the mice had similar bone mass. The effect on bone showed up early only in those that were given antibiotics. In later experiments that I describe below, we also found bone effects, some of them lifelong. Again, this was not specific to a single antibiotic. If it was, one might think of it as a side effect of that one drug. But it was present across all the antibiotics tested. This work supports the idea that, in addition to better nutrition and clean water, antibiotics may be part of the explanation of why people are taller than ever.

We now had evidence that STAT changes early development but still did not understand how it happened. How did adding antibiotics to the water cause these developmental effects? What made the animals fatter and built up their bones earlier? We suspected that the drugs changed the composition of the intestinal resident microbes, so that is where we looked first by examining mouse poop. Fecal pellets represent the end product of everything that happens in the intestine and could be collected every day from each mouse. The pellets gave us a standard material to compare in the same mouse over time, and across the mice that were exposed to different antibiotics or not, and varied diets.

We also studied material from an upper region of the colon called the cecum after sacrificing the animals. Cecal content was important to our study because it showed us which microbes were present and active in the body, not only after elimination in the feces. Because it had to be removed surgically, we could collect it only once, when the mice were killed. Most intestinal content of mice and humans, whether in the colon or in the poop, is undigested dietary fibers, water, and bacteria; the DNA present is nearly all bacterial. We performed what's called a universal bacterial 16S ribosomal RNA assay to learn more.

All bacteria share a gene that encodes 16S rRNA, which they need to make proteins. Although all bacteria have 16S rRNA genes, the exact DNA sequence substantially differs among bacterial species. The form in *E. coli* differs from that in *Staph.* So, by using the universal technique, followed by sequencing of the DNA products, we can take a census of "who is there." It is similar to taking a census in New York or Chicago and asking how many teachers, lawyers, police officers, and schoolchildren live there. In this case, we are asking how many clostridia, bacteroides, streptococci, and so on are present, down to thousands of individual bacterial species. Based on the results of our census, we were able to address a number of important questions.

First, does the STAT treatment alter bacterial diversity? In other words, are the resident microbes of the antibiotic-treated mice as diverse as those of the control mice? Although both samples might be expected to have a lot of schoolteachers, students, and police officers, because they are common, will they have actuaries and piano tuners (rare professions) or had those dropped out?

We found that STAT, possibly because it is low dose, had no obvious effects on bacterial diversity. The same number of "professions" were present in the STAT-exposed and control specimens.

But what happens to the composition—the relative proportions of teachers, police officers, and so on—with STAT? We can take a census of who is there. For example, we would expect that the distribution of these professions in New York and Chicago would be closer in composition with each other than either would be with Delhi or Beijing. This is a model of what we find in the gut microbiome.

This is where things got interesting. STAT changed the composition of the intestinal microbial population, whether we examined the fecal pellets or the cecal contents. We expected that usual antibiotic exposures would change the mix, but we didn't know whether very low doses of STAT would do the same. We found that they did.

But did they change the functions of the bacteria? The answer is yes. Most of the food you eat is digested and absorbed in your small intestine. Residual food that reaches your large intestine is mostly indigestible. But here your bacteria come to the rescue. Recall that certain microbes in the colon digest this material and produce what are called short-chain fatty acids (SCFA), which are absorbed in the colon. These SFCA represent 5–15 percent of the calories you take in every day. If your microbes were more efficient at extracting calories from this "indigestible" food, then you would be better nourished. You might get fatter.

We measured SFCA levels in cecal contents and found that they were significantly greater in the STAT mice than in the controls. That meant that STAT mice were getting more calories early in life from their microbes, just as their tissues were developing.

We next zeroed in on the liver, the body's main metabolic factory. It transforms the food absorbed in the intestinal tract, including the SCFA, into useful products, including proteins, energy sources such as sugars and starches, and energy-storage molecules such as fat. We compared the genes expressed in the livers of STAT mice to those of the control mice.

We were right on target. The liver in STAT mice up-regulated the genes needed to make and transport more fat out to the periphery—the blubbery layers of fat animals. We knew that the STAT mice were putting down more fat and that it had to come from somewhere. The liver made sense. It is strategically interposed between the intestinal tract, where energy is acquired or generated, and the adipose tissue, where fat is stored.

■ ■ ■

Our next experiment, planned and carried out by Laurie, examined in more detail what happens when mice get antibiotics (we chose penicillin) very early in life. In Ilseung's experiment, animals got the drugs when they were weaned, about twenty-four days after birth. This is equivalent to at least twelve months for a human baby. Now Laurie gave antibiotics to the mothers during their pregnancy, so that their microbes, including those in the vagina, were altered from the get-go. The infant mice began life exposed to an altered microbiome, and we continued to give them antibiotics. As we predicted, the mice exposed at birth grew more than those exposed at day twenty-four. That became our standard way to conduct experiments.

Next, Laurie conducted an experiment that zeroed in on *when* the mice begin to get fat. Mice grow rapidly from the time they are born. Would they be putting on the extra fat in their youth or did it take a while? The results of the experiment were clear. In males we found a difference from controls at sixteen weeks, and in females fat showed up at twenty weeks (middle age for a mouse). But in both sexes, once it was there, the increased fat persisted for their entire life span.

Subsequently Laurie looked at which species of bacteria were prevalent in these young mice. At four weeks, control animals were dominated by *Lactobacillus*, the bacteria originating in their mother's vagina. This was expected because the animals had just finished nursing, a time when, in both mice and humans, lactobacilli dominate.

But in the STAT group most of the lactobacilli were gone, replaced by other groups of bacteria. Since the changes in body composition were detected after sixteen weeks and the resident microbes were different at four weeks, we had a critical observation: changes in the microbiome preceded the changes in body composition.

Some elegant work done by my longtime friend and colleague Jeff Gordon at Washington University in St. Louis adds insight to our findings. Jeff has been a giant in the field of microbiome science, building on his years of research on how the gastrointestinal tract develops and functions. Jeff's group studied mice with a deletion in the gene responsible for the production of leptin, the "feed me" hormone that helps regulate appetite and helps the brain decide whether to store or use energy. Leptin-deficient mice, called *ob/ob* mice, become markedly obese. Jeff and his colleagues asked whether the resident microbes of the *ob/ob* mice differ from those of their normal littermates. The answer was yes. Each type of mouse had different microbial populations in its guts.

Then Jeff asked if the microbes performed different metabolic roles. He transferred intestinal contents from the obese *ob/ob* and the normal mice into germ-free mice. These mice have thinner intestinal walls with fewer cells and do not gain as much weight. But when they are conventionalized and get microbes back, how well do they grow? Jeff's finding, which made news around the world, is that resident microbes taken from obese mice caused the recipient mice to put on fat at an accelerated rate compared to the mice that received microbes from the normal-weight mouse donor.

But here is something to consider: the mice in Jeff's experiments had a genetic defect that made them obese to begin with. That was the cause of the obesity; the change in the microbial populations was secondary. Although Jeff's team had beautifully characterized the consequences of obesity on the microbes and their functions, I did not think they were addressing the root cause of obesity. Moreover, germ-free mice, which provide an elegant system for testing specific hypotheses about immunity and metabolism, are completely artificial. Yet although there are no natural germ-free mice or humans, we still can learn much about the fundamental principles of host-microbial interactions.

My own view was that antibiotic-induced perturbations in resident microbes early in life in relatively normal individuals might be the primary events that change host metabolism. (It would be about two years until we had more definite proof.)

Next, we asked what happens if we combine STAT with a high-fat diet. As we all know, our children's diets have gotten a lot richer in recent decades, whether from sugary drinks or from high-fat foods. They are taking in more calories on average than kids did one and two generations ago. We know that mice get fatter on a calorically rich diet, but would STAT increase or decrease the trend, or would it just be neutral?

Laurie called this experiment FatSTAT, and again the results were exciting. As we expected, mice on the high-fat diet got bigger than animals on normal chow. But adding antibiotics made a significant difference. We had mimicked the manner in which modern farmers are raising their livestock. Males on the combination (fat diet and antibiotic) were about 10 percent bigger still, having gained both muscle and fat. But the most striking differences were in the amount of body fat: with the combination, males had about 25 percent more, but females had an astounding 100 percent more. The females on the high-fat diet gained about 5 grams of fat, whereas those on the fat diet and antibiotic gained 10 grams. They doubled their body fat. That's a lot, considering their total body weight was 20–30 grams.

Thus antibiotics had an effect, high-fat diet had an effect, but together they were more than additive; they were synergistic. For the female mice, the antibiotic exposure was the switch that converted more of those extra calories in the diet to fat, while the males grew more in terms of both muscle and fat. We do not yet know the reason for these sex differences. Nevertheless, the observations are consistent with the idea that the modern high-calorie diet alone is insufficient to explain the obesity epidemic and that antibiotics could be contributing.

We asked another simple question that Laurie's thesis committee suggested to us. Up to this point, we were keeping the animals on

STAT for their entire lives. Would a few weeks of antibiotic treatment be enough for the weight gains to persist? This was a question
important for our children's future. If the weight gain happens only
after long-term treatment, then maybe this isn't relevant to our kids.
Very few get lifelong antibiotics. But if short-term exposures cause
the problem, this may be a way to explain our current epidemic.
Most children are getting relatively short exposures of antibiotics for
their ear and respiratory infections, especially early in life.

In March 2011, Laurie began the DuraSTAT experiments, so
named because we were testing the durability of a brief antibiotic
exposure to produce an effect. She divided the mice into four groups:
no antibiotics, which was the control group; STAT for only four
weeks and then stop; STAT for eight weeks and then stop; or STAT
for the duration of the experiment. All of the mice were put on a high-
fat diet at six weeks to bring out any differences. Laurie focused on
females because of the results of our FatSTAT study.

Mice getting continuous antibiotics for the duration of the experiment gained weight compared to the controls, just as expected. But
the effects of getting antibiotics for four weeks or eight weeks were
the same as for twenty-eight weeks. The mice receiving the penicillin
gained 10–15 percent more in total weight and 30–60 percent in fat
compared to the antibiotic-free control mice. In other words, exposure
to STAT early in life was sufficient for a lifelong effect; the development of the mice changed. Although the results in DuraSTAT were
not identical to those of FatSTAT, neither were the experimental
conditions. So the experiments are not directly comparable. The relevant comparisons are within each experiment. This is an important
issue in science, where investigators have gone astray by comparing
the effects in one experiment with those in a different one; conditions
change in ways that often are not being measured. But for us, the trends
were exactly the same: early-life STAT permanently changed development in these mice.

Next we decided to study the microbiome itself. Laurie had been
faithfully collecting the tiny fecal pellets, often once a day, from

every mouse. She had thousands of little plastic test tubes in white boxes, one pellet per tube, one hundred tubes per box. It would take about eighteen thousand pellets to make a pound. They were worth more than their weight in gold because of the secrets they carried.

Laurie sequenced hundreds of specimens to determine their DNA compositions and to learn about the structure of their microbial communities (let's say again assessing the ratio of teachers to police officers but now in much greater detail), including tax lawyers, taxi drivers, and taxidermists.

First she looked at samples from newly weaned three-week-old mice that had been given penicillin and compared them with samples from control mice that did not get the drug. Although the community structures of the two groups overlapped some, they were clearly different. This was exactly as we expected: antibiotics affect the structure of the microbial community in the intestinal tract.

Then we looked at pellets obtained eight weeks into the study. Now there were essentially three groups of mice: the controls (no antibiotics), the mice still receiving antibiotics, and those who stopped antibiotics after four weeks and drank plain water for the next four weeks. As expected, the microbial community structure of controls and mice on continuous antibiotic were even more dissimilar than they were at three weeks. Antibiotics work. But the microbial community structures of the mice that had stopped taking antibiotics now looked just like those of the controls; they nearly overlapped. This means that the major effects of four weeks of antibiotics on the community structure were just transient. This was very clear. Yet, remember, these mice got just as fat as the others, which suggests that a brief exposure to antibiotics early in life, which causes an early perturbation of resident microbes, can lead to a lifelong effect. And the perturbation need not be permanent.

This is a key finding. I believe that it is the paradigm for what is happening to our children. Disturbing the microbes of mice during this critical early window is sufficient to change the course of their development. This was the experiment that proved to me that antibiotics have the potential to change development. And, of course, development is multidimensional: it is metabolic, as we were studying in the mice, but it is also immunologic and cognitive. As babies grow, while they are sleeping and dreaming, the context of their later development is being formed in partnership with their ancient microbes. Even transient perturbations at that critical time can make a big difference.

But we are scientists, and we need to keep extending the story, learning the details, and finding the mechanisms. We need to answer the seemingly simple question: How does it work? What is so important about antibiotic exposure? Is it just its effects on the microbes, or is it because the penicillin had other effects on the body, directly interacting with the tissues of the mouse, irrespective of the effects on the microbes? As with many prior experiments, including those conducted by Jeff Gordon, we would attempt to answer the question by transferring microbes between mice.

Recall our earlier question: Was weight gain a direct effect of the antibiotics or was it a result of how the antibiotics affected resident microbes? We presumed it was the microbes, but presumption is not proof. To find out, we needed to transfer the STAT or the control microbes into a neutral situation and then observe whether there were differences in the recipients. Like Jeff, we chose to study the effects in germ-free mice.

We bought fifteen germ-free female mice, and in late August 2011 they arrived in three plastic bubbles, five to a bubble, newly weaned at three weeks old. The company told us that we could maintain them in the bubble for up to seventy-two hours, enough time to start our experiment. We called it TransSTAT because we were transferring the STAT-affected microbiota to recipient mice.

Laurie chose six eighteen-week-old mice from her DuraSTAT experiment: three controls and three on continuous antibiotics. She collected cecal contents from each mouse and pooled them into two groups, one from controls, one from STAT mice. Calling on her extensive background in bacteriology, Laurie took special steps to preserve the viability of the microbes, some of which are so sensitive to oxygen that even a brief exposure to air kills them. Then she introduced the cecal matter into the stomach of each germ-free mouse. Seven received pooled cecal contents from controls, and eight received the cecal contents from STAT mice. To you and me the introduction of cecal contents into the stomach seems particularly unappetizing, but mice are coprophagic, meaning that they regularly eat their own feces as well as the feces of cohoused mice.

Now the mice were no longer germ-free. They had been "conventionalized" and could begin the next phase of their lives with their own residential microbes. We followed them for five weeks, obtaining frequent fecal samples and taking measurements, including DEXA scans, four times on each mouse. None of the mice received any antibiotics. All were raised identically, differing only in which microbes they received.

As expected, all of the mice gained weight since they were still growing. However, the mice given the STAT microbes gained more weight and had more fat than the mice that were fed the control microbes. Nor were the effects small. The STAT recipients gained about 10 percent more weight and about 40 percent more fat than the control recipients did.

With this experiment, Laurie proved that the STAT-induced changes in development were transferrable by altered microbes alone.

■ ■ ■

STAT showed us what happens on the farm. But I am mostly interested in human children. When they get antibiotics, the dosage is rarely continuous. Rather, as discussed earlier, they receive short courses, usually five to ten days, depending on the problem (ear infection, bronchitis, sore throat) and on the doctor.

I wanted to see whether short pulses of antibiotics would affect weight gain and fat. Thus came the new model, which we called PAT, for pulsed antibiotic treatment. Instead of low doses, mice got the antibiotics just like children do, full therapeutic doses for just a few days in several pulses.

We chose amoxicillin and tylosin, which together represent more than 80 percent of all the antibiotics prescribed for American children. We then selected four groups of mice: controls that did not get antibiotics, a group given amoxicillin in three pulses, a group given tylosin in three pulses, and—thinking there might be an additive effect—a mixed group that alternated tylosin, amoxicillin, tylosin for their three pulses.

To get the drugs into the baby mice as early as possible, Yael bred adult females and put antibiotics (or not, for the controls) into their water ten days after they gave birth. We guessed the drugs would be absorbed into the bloodstreams of the moms and get into their milk and thus affect the microbes in their pups—an assumption that proved to be correct.

The first PAT exposure occurred when the babies were ten to fourteen days old. At twenty-eight days old, after they were weaned and on their own, and again at thirty-seven days old, they got three-day pulses in their drinking water. On day forty-one, we switched all the mice to a high-fat diet, so we could enhance differences induced by antibiotics. All the mice we studied were females, because the breeding resulted in a more regular group than the males.

By day twenty-eight, all the PAT mice were growing significantly faster than the controls. We performed analyses on fat, bone, and muscle for the next 150 days of life, which took the mice well into middle age and early old age. PAT mice showed more muscle mass than the controls but not much difference in fat mass. Bone was a different matter. The PAT mice that received amoxicillin showed increased bone area and mineral content for the duration of the experiment. Perhaps the effect was permanent because they received the drugs so early in life. And since amoxicillin is the most frequently prescribed drug in childhood, I can only wonder if that's the drug that most promotes the recent increases in human height.

Yael had collected more than three thousand fecal pellets from the mice and, for each specimen, knew which mouse it came from, on which day, and which treatment the animal got. With help from colleagues at Washington University in St. Louis we took a close look at their DNA. We wanted to know how our treatments had affected each animal's intestinal microbial diversity.

We found that moms had an average of 800 species in their fecal pellets. After one pulse, pups in the control group were like the moms. Those in the amoxicillin group had about 700 species. But mice in the tylosin and mixture group had only 200 species. In other words, the one course of antibiotics had caused the suppression or disappearance of about two-thirds of the usual bacteria in their feces. We saw a similar effect with amoxicillin, but it was much milder.

Now, after the three courses were over, we wondered whether the richness and biodiversity of the bacterial species would bounce back. It mostly did for amoxicillin, which is a relatively mild drug. But in the mice that had received the tylosin, the diversity never went back to normal, even months after the last antibiotic dose. Tylosin had permanently suppressed or wiped out a proportion of the organisms passed on to them by their mother.

We also measured the so-called evenness of microbial diversity. If it's high, it means most species are found in roughly equivalent numbers. If low, only one or a few species dominate. In a human society, we could compare peacetime, when many different professions are well represented, and wartime, when there is a huge increase in the number of soldiers and corresponding decreases in all other professional groups. In war, the professional structure of society changes markedly. Tylosin treatment gave us the wartime equivalent with low evenness. PAT was causing permanent changes to the structure of the microbial community early in life, just as the mice were developing.

■ ■ ■

All told, our STAT and PAT experiments built a strong story that early-life antibiotics change the development of mice through their effects on resident microbes. But mice are not humans. We wanted to know if anyone was trying to link obesity with antibiotic use in young children. Despite a profusion of published studies on childhood obesity—including investigations into birth weight, time spent watching TV, amount of exercise, exact details on every dietary nuance—as well as some big studies now under way, no one to our knowledge had ever asked about antibiotics.

Then my colleagues Drs. Leo Trasande and Jan Blustein heard about the Avon Longitudinal Study of Parents and Children (ALSPAC) study in Britain. Beginning in 1991 more than 14,500 pregnant women in the Avon Health District were recruited for a study. Their children, enrolled at birth, became a cohort that was studied for the next fifteen years. We were particularly interested in the kids who became overweight or obese.

Luckily for us, there was exactly one question on the questionnaire each parent filled out periodically. As part of a survey of the drugs to which the kids were exposed, they asked: "Did your child use any antibiotics in the preceding period?" It was asked when the kids were six, fifteen, and twenty-four months of age.

Nearly a third had received antibiotics in the first six months of life. By age two, three-quarters had been treated. Did the antibiotics make any difference? The calculations were complex, and it took an excellent statistician like Leo to make sense of them. Leo had to examine the effects of antibiotics while controlling for such factors as the baby's initial weight and the mom's weight and whether the baby had been breast- or bottle-fed and for how long.

The upshot: children who received antibiotics in the first six months of life became fatter. We weren't surprised; the earlier in life, the stronger the effect on farm animals. Laurie had shown that early-life dosing was more important in the mice and that if we had to guess which time period would be most important for the development of human babies, it would be the first months of life.

So on the farm, in our mouse experiments, and in an epidemiologic study of human children, there was consistent evidence that early-life exposure to antibiotics could change development leading to larger size and more fat. We are testing more variations in mice, but the story continues to hang true, as we begin to fill in more and more details of the plot and the characters.

■ ■ ■

After the first epidemiologic study, Jan and Leo used ALSPAC to look at methods of birth delivery. Using parallel statistical analyses, they found that C-section births also were associated more with obesity. This was one of several studies concerning U.S., Canadian, Brazilian, and now English children published in 2011–13 that examined the same question. There were differences in design and findings among all of the studies—for example, we found that nearly all of the effect occurred when the mom already was overweight. However, in each studied population C-section was associated with worse outcomes but the other (confounding) factors may also contribute to the risk. No one had ever made the connection between C-sections and childhood obesity. Maybe the informed consent form that a woman signs before she undergoes a C-section will say in the future: "One of the risks of this procedure is that your baby has an increased risk of becoming obese, and developing celiac disease, asthma, allergies . . ."

I list these other conditions here, because there have been convincing studies showing the relationship of C-section with these other modern plagues. Now that we know how medical interventions affect development and could lead to these diseases, perhaps we can find ways to prevent and treat them. But first, let's examine them more closely.

14.

MODERN PLAGUES REVISITED

In 1974, when Kathy was thirteen, she had to get a routine physical before going to summer camp. I knew her family well. She was so full of energy, no one expected what was about to happen. The doctor called her mother to report that Kathy's urine had sugar in it. "She has diabetes," he said. "It appears to be mild, but you will need to watch her very carefully." Kathy's grandfather had developed it in his forties and died in his early fifties. Still, this was a shock.

At first, Kathy was lucky. Many children diagnosed with diabetes fall desperately ill. They lose weight rapidly, wet their beds, are constantly thirsty, and feel painfully exhausted. But Kathy had no such symptoms. She was an athletic, healthy girl with thin brown hair, brown eyes, glasses—as normal as can be. For that first year, she was able to control her diabetes by watching her diet. But being a young teenager, Kathy resented the restrictions suddenly placed on her. She would rebel and have ice cream cones with friends after school and deliberately ignore instructions from her diabetes nurse.

A year later, Kathy's blood sugar rose to dangerous levels and she had to take insulin shots every day. This is when she developed a lifelong anger toward her illness. It wasn't fair that she couldn't eat whatever she wanted. It wasn't right that she had to live life differently. Admonitions to keep her disease under "tight control" went unheeded. Soon Kathy needed insulin twice a day. A few times when her blood sugar dipped too low, she had to be hospitalized.

Despite the ups and downs, Kathy lived her life fully, showing enormous courage, willful strength, and a generous spirit. She graduated from college, became a social worker, married, and had a daughter when she was twenty-five. The diabetes complicated her pregnancy. She tried an insulin pump, but it didn't work well for her, and she never tried again. After she gave birth, her sugar levels settled back down for a while, but eventually the roller coaster of struggling to control her blood sugar resumed. Kathy continued to occasionally binge on forbidden foods. She did not exercise. She played around with her insulin levels.

Over the years, the diabetes took its toll. Kathy lost sensation in her feet and the tendons in her hands began to contract, contorting her fingers. When she was thirty-five, her daughter, then nine, also became diabetic and was started on insulin injections. Doctors blamed the condition on a genetic predisposition, which made Kathy feel horribly guilty.

Nevertheless, Kathy persevered. Well into her forties, she was very much her own woman. She divorced, remarried, adopted a son, and lived life on her terms, not the disease's. But then her kidneys began to fail. She was put on the kidney transplant list. She had a heart attack when she was forty-six. Her diabetes became even more difficult to control, with lots of episodes of low blood sugar. She grew very thin. One day in 2011, Kathy felt confused. The next moment she was in a coma, and one week later she was dead. She just missed her fiftieth birthday.

Type 1 or juvenile diabetes is an autoimmune disease in which types of T-cells—immune cells that recognize foreign proteins called antigens—turn against the body's own proteins. In this instance,

T-cells attack the pancreas and destroy the islet cells that make insulin. The disease can occur at any age but is most commonly diagnosed from infancy to the late thirties. (In contrast, adult onset or Type 2 diabetes is a condition of insulin resistance in which the body's cells fail to respond to insulin properly. It is linked to obesity and tends to occur later in life.)

Insulin is the key that allows the major form of sugar circulating in the blood, called glucose, to enter and nourish cells throughout the body. When Kathy's islet cells were destroyed, her insulin production collapsed. Without insulin, her tissues were starving, even though her bloodstream was full of glucose. The sugar could not get into her cells. Because her kidneys could not filter the excess sugar, it spilled out of her body in her urine so often that she became dehydrated. Essentially she was peeing out the calories that her body could not absorb.

Once Kathy began taking insulin shots, she was able to keep her blood sugar in a more normal range. But danger lurked constantly. If she took too much insulin, her blood sugar could fall dangerously low. She might get shaky, sweaty, or even pass out. When her blood sugar remained too high over the long term, her heart, blood vessels, nerves, skin, and kidneys all were damaged.

I'm telling this story not to convince you that diabetes is a terrible disease (although it is) but rather to sound an alarm about its sudden rise to epidemic proportions. Today, the rate of Type I diabetes is doubling every twenty years all over the developed world; moreover, children are coming down with the disease at younger ages. When Kathy was diagnosed, the average age of onset was about nine. That means that at nine essentially all of the insulin-producing cells in a patient's pancreas were already gone, which means that the process had actually begun years earlier. But now, the average age of onset is around six, and some children are developing the illness when they are only two or three. This means that for some children, their islet cells are disappearing before their second birthdays.

Of course, there are many hypotheses about why this increase is happening. While several genes are known to predispose children to the disease, and such genes may have been carried by Kathy's grandfather and passed down (sometimes entire generations are skipped), recent research has focused on environmental factors that might be triggering it. These include our old friend the hygiene hypothesis, viruses, vitamin D deficiency, and antibodies resulting from drinking cow's milk.

As I studied the literature, I found other markers for risk of the disease. Juvenile diabetes is more likely to develop in babies born by C-section, in boys who are tall, and in babies who gain weight more rapidly in the first year of life. Each of these observations suggested to me that perturbation of our resident microbes very early in life could be a contributing factor.

In March 2011, at a meeting of the Human Microbiome Project, I met Jessica Dunne, a very dedicated program officer from the Juvenile Diabetes Research Foundation, who invited me to give a talk at the organization's headquarters in New York. She had heard me speak about our work on antibiotics and obesity and was curious about our thoughts on diabetes.

I was lucky, because at the time I had begun to work with Alexandra Livanos, an NYU medical student who was interested in how pancreatic inflammation affects the microbiome. I suggested to Ali that she change the focus of her project from the pancreas in general to Type 1 diabetes in particular. It wasn't a huge shift in target—we still were studying damage to the pancreas—but it was a big change in terms of what we were looking for and how we would approach it.

By July, Ali had begun to study the effects of early-life antibiotics in a mouse strain (the NOD mouse) that spontaneously develops a disease strongly resembling Type 1 diabetes in humans. We had a hypothesis in mind. Studies show that various treatments can delay diabetes. But could anything accelerate it? Our idea was that antibiotics would speed up the onset and severity of the disease.

In the meantime, in applying for funding from the diabetes

foundation, I proposed that we study diabetic mice exposed to either the subtherapeutic regime (STAT) or the pulsed therapeutic doses of antibiotics (PAT). Happily we were funded but got only half of the support we requested. Limited for funds, the foundation said that we should focus on STAT rather than PAT, because our preliminary data on obesity looked more promising. Fortunately, I had some research money in reserve that allowed us to include both in the experiments.

The work is in progress as I write this, but Ali has already presented some of her preliminary results at scientific meetings. Her studies show that the disease comes on faster after PAT, but so far the effect is seen only in males. Even before the diabetes developed, the pancreases in antibiotic-exposed mice looked terrible, with angry immune and inflammatory cells tearing up the insulin-producing islets. Ali also found that intestinal immune cells are altered by the drugs, again before the onset of diabetes. This is evidence that an abnormal interaction in the intestine precedes the destruction in the pancreas. Very recently, Ali showed that PAT significantly alters the composition of the resident microbes well before the mice develop early diabetes and certain organisms stand out as potentially being protective. Interestingly, all of the PAT effects were stronger than STAT, so we had made the right decision to pursue both research avenues.

Thus juvenile diabetes is another disease in which early exposure to antibiotics could be playing a causative role, at least in accelerating the disease. Once again, mice are not humans, but these early findings are consistent with our notion of the risk of perturbing the early-life microbiota, in this case, while the immune system is developing. We are in the middle of follow-up experiments to better understand the disease mechanisms involved, and we have teamed up with investigators in Massachusetts, Florida, North Carolina, and Sweden to extend our inquiries. But so far, in mice—at least in the males—we have evidence that early-life antibiotic exposures are increasing and worsening the risk of developing Type I diabetes, in terms of both the number of individuals affected and the age of onset, just as we had hypothesized.

■ ■ ■

My daughter Genia was born in 1983, and like many kids she had a lot of ear infections. In those days, pediatricians often advised putting tubes into a child's ears to treat the infections, but as a doctor I didn't like that idea because it could permanently scar her eardrums. Her doctor agreed with me, so until she was six or seven Genia received many courses of antibiotics for days or weeks on end. It was mostly amoxicillin, the pink liquid with the ultrasweet odor and taste. Her story is not unique.

As Genia was growing up, she developed a little asthma and some food allergies, including a severe reaction to the skin of mangoes. But overall, she seemed to outgrow her asthma, which was mild and, by avoiding mangoes, she had no further problems.

Genia has since become what my late mother would have called a great humanitarian. Since her teenage years, she has been going to Latin America, working, studying, helping disadvantaged people, and traveling and exploring along the way. Not surprisingly, based on how she traveled, where she stayed, and what she ate and drank, she had multiple episodes of diarrhea, which travelers and doctors call *turista* or *Montezuma's revenge*. Sometimes these bouts lasted a few weeks. A few times she had a particularly uncomfortable intestinal infection caused by a protozoan called Giardia. Such infections are commonly treated with the antibiotic metronidazole (also sold as Flagyl). Metronidazole, which is often used to treat intestinal infections, not only targets Giardia but also has major, broad effects on the resident bacteria of the gut. Genia took four courses of metronidazole in 2008 and 2009, but her abdominal pain only became more frequent.

After working in Ecuador in 2009 and taking another course of metronidazole, Genia developed severe abdominal pain and continuing diarrhea. Her symptoms persisted for months. Blood tests indicated that she was anemic and that her body wasn't absorbing certain vitamins very well. By then, she had returned to Boston to study law. One night, Genia's symptoms were so bad that she went to the emergency room at Massachusetts General Hospital. The doctors there thought that she might have acute appendicitis, but fortunately her symptoms improved before they had time to operate.

I was distraught about her condition. I have colleagues all over the world who are specialists in abdominal problems, the diseases of travelers. I enlisted several truly outstanding doctors to evaluate her, but neither they nor I could find the cause of her troubles. She took blood tests for celiac disease, an intestinal malady that produces similar symptoms, but the assays were normal.

With celiac disease (the name derives from the Greek word for *hollow* as in *bowel*) people are allergic to the main protein in wheat (which also is present in barley and rye), called *gluten*. Even eating a tiny amount of gluten can trigger an immune reaction that attacks healthy cells lining the small intestine. In other words, the immune system treats gluten as a deadly invader, not as food. Symptoms include abdominal pain, diarrhea, bloating, and fatigue. Even if a person manages to avoid gluten for months, the symptoms can come back immediately on reexposure.

The incidence of celiac disease has skyrocketed in recent decades, more than quadrupling since 1950.

In 2009, Genia underwent endoscopy of her upper intestine, including a couple of biopsies to look for celiac disease, but again the tests came back normal. Meanwhile, her symptoms continued, lasting for more than a year. She was miserable.

A friend suggested that she might have a food allergy, and in May 2010 she went to see Dr. Bernard "Rardi" Feigenbaum, a colleague of mine who is a terrific allergist. He thought she might have celiac disease, despite negative test results. Sometimes people with otherwise classic symptoms test normal. This made sense to me because I have found that people often don't have the textbook version of any particular disease. I know from my own experience that the chapters in the textbooks that I both write and edit contain general rules that try to cover the most important situations. But forms of illness are legion, and understanding variation is central to what good doctors must do. (That is one of the biggest dangers of cookbook medicine: we stop thinking, searching, analyzing, because we follow the guidelines.)

Rardi suggested that Genia start a gluten-free diet, to see if it helped. She did, and almost immediately her symptoms went away. For the first time in months, she didn't have abdominal pain. Unfortunately, people with celiac disease face a serious problem: gluten is everywhere. One night, the sharp pain came back. She realized that she had eaten soy sauce at a restaurant, which is when she discovered that soy sauce often contains gluten.

After that she avoided gluten and felt great for a month. Then one day as she was traveling on the interstate highway back to Boston and stopped for fast food, she ordered french fries. Again, within an hour, abdominal pain struck her as severely as it had when they thought her appendix might burst. Later she learned that those french fries contained gluten.

Nowadays Genia is scrupulous about what she eats and has been almost entirely free of symptoms since the french fries episode. As a scientist, I can't prove that Genia has celiac disease, but her symptoms are certainly consistent, and a more recent blood test showed for the first time that she has elevated levels of gluten antibodies. What piques my interest is all of those courses of amoxicillin that we gave her as a child, and then the courses of metronidazole for her "Giardia" infections. It suggests to me that she had major early-life disruptions in her resident microbes that led to her asthma and mango allergies. Later on, metronidazole was the coup de grâce, eliminating some population of microbes in her gut that instructed her immune cells to suppress certain allergies, including gluten.

Not long ago, a group of colleagues whose work focuses on celiac disease asked me to help them analyze some data collected in Sweden. Drs. Karl Marild and Jonas Ludvigsson had obtained records on thousands of people diagnosed with celiac disease, people with conditions similar to celiac disease (I'll call it almost-celiac, perhaps like

Genia's case), and healthy controls. Plus they were able to access nation-wide pharmacy records.

The main finding: people who recently had developed celiac disease were about 40 percent more likely to have been prescribed antibiotics in the preceding months compared to those who didn't. This was true for people who definitely had celiac disease, for those who probably had celiac disease (60–90 percent more likely), for men, for women, for people of all ages, and for essentially every antibiotic examined individually. For those who were prescribed more courses, the risk was stronger. As with the diabetes studies, this kind of consistency was very important; it was not just a single finding. Most interesting to me is that metronidazole (the same drug that Genia was given repeatedly), which has major effects on gut bacteria, had the greatest association with celiac disease. People who were prescribed it ran more than twice the risk of getting celiac disease compared to those who were not recently prescribed any antibiotics.

To be sure, these studies show only an association between antibiotic use and celiac disease. We don't even know whether those who were prescribed antibiotics took their medication, but we assume that people generally did, whether or not they later developed celiac disease. We also can't prove what is called the direction of causality. One possibility is direct causality: exposure to antibiotics brings out the tendency for celiac disease. Another possibility is reverse causality: people already have celiac disease, and their doctors give them antibiotics to treat their symptoms without realizing what they have. At this point, we cannot distinguish which is correct, but clearly the first hypothesis is consistent with our work and with Genia's case.

I was invited to join another analysis of celiac disease by the same group of experts who now were led by Dr. Ben Lebwohl at Columbia University. The question was: Is there any relationship between the presence of *H. pylori* in the stomach and celiac disease? We know that *H. pylori* is disappearing just as celiac disease is becoming more common. Could this suggest a protective relationship? *H. pylori*, when present, is almost always acquired very early in life before celiac disease develops. We also know that *H. pylori* helps suppress immune and allergic responses through its recruitment of T-reg cells, the cells that tone down and turn off immune reactions. So could the extinction of *H. pylori* be contributing to this modern plague as well?

To find out, the Columbia group, working with a team of pathologists at a large national reference laboratory in Texas, investigated more than 136,000 people who had undergone upper gastrointestinal endoscopy for a variety of reasons. As part of their routine analyses, the pathologists looked for *H. pylori* in stomach biopsies and characterized inflammation in the duodenum. Since signs of celiac disease can be detected microscopically in the duodenum, Ben and his colleagues realized that they had a way to link celiac disease to the presence of *H. pylori* in the stomach. Would people who had *H. pylori* be more likely to have celiac disease or less likely, or would there be no association?

Because the study was performed in the United States, the overall rate of *H. pylori* was very low. Nevertheless the rate in patients who had signs of celiac disease was only 4.4 percent as opposed to 8.8 percent in those without. Because of the large number of subjects, we asked whether the same relationship was true of the subjects in the thirty-seven individual states from which the specimens had come, and indeed it was in every single state studied. The reciprocal relationship held in both men and women as well as in each age group studied. These consistencies suggest biological significance.

It is plausible that celiac disease is increasing because the microbes that protect against allergic responses are disappearing. Both stomach bacteria (*H. pylori*) and intestinal bacteria (those susceptible to metronidazole and/or other antibiotics) may provide some protection against celiac disease. People who have *H. pylori* in their stomachs can still develop celiac disease; it just may be less likely. Moreover, people born by C-section also face an increased risk. With this knowledge, we might one day be able to find the individually protective organisms and either prevent their loss or give them back to prevent celiac disease or to treat it.

Another condition to consider in the light of missing microbial diversity is what is called inflammatory bowel disease (IBD), a group of chronic, relapsing disorders of the intestine. IBD manifests in two main types, ulcerative colitis and Crohn's disease, which partially overlap but have different pathology.

Ulcerative colitis affects just the colon and is most often limited to the most superficial layers of the bowel wall. People often bleed from their rectum, have severe diarrhea, lose weight, and become anemic. The disease can ruin their lives. And to make matters worse, the longer they have it, the greater their risk of cancer. One of my closest friends had ulcerative colitis until about ten years ago, when he had his colon removed. It was the right thing to do. His disease was out of control, and after more than thirty years of having an inflamed colon, his risk of colon cancer was much too high. He now has a bag, called a colostomy, in place of a colon. Although not ideal, he has added a new repertoire to his jokes. But he doesn't have those bouts of illness anymore, and he can hike up mountains faster than I can.

Crohn's disease can involve the entire GI tract, where patches of inflammation appear and where the resulting scarring, called fibrosis, leads to intestinal obstructions. Although ulcerative colitis has been recognized for a long time, Crohn's disease was first described in 1932 by Dr. Burrill Crohn, a New York physician. Is Crohn's disease new to the twentieth century, or had it been missed previously? We don't know the answer to that, nor do we know what causes it. But we do know that the incidence has been rising in recent years all over the developed world and even in developing countries that are becoming more industrialized.

It's clear that intestinal microbes are involved in IBD because nearly every mouse model of the disease requires their presence for colitis to develop. The illnesses wax and wane, and antibiotics help people get through their crises. Still, this doesn't tell us whether the bacteria are of primary importance or secondary, but they clearly are involved at some level. The more important question is: Why is the incidence of IBD increasing?

In 2011 a group of Danish investigators reported on their examination of pharmacy records for all 577,627 children born in Denmark as singletons (not as twins) between 1995 and 2003 and their assessment of their risk of developing IBD at an early age. The children were followed for an average of nearly six years, representing more than 3 million person-years of follow-up. A study of this scale provides an opportunity to observe uncommon events.

One hundred and seventeen children developed IBD. They had their first contact with the medical system—in a clinic, an emergency room, or a hospital—when they were on average three years and five months old. These are very early cases of IBD; the peak generally comes later in life, and, predictably, there will be many more afflicted persons. Still, the researchers could identify what these children were exposed to before they got sick and seek correlations. Compared to healthy children, those who developed early IBD were 84 percent more likely to have received antibiotics. Furthermore, children who had taken antibiotics had more than triple the risk of developing Crohn's disease than those who were antibiotic-free. The more often they took antibiotics, the higher the risk. The investigators calculated that each course of antibiotics was associated with an 18 percent increased risk of developing Crohn's disease. The children who had received seven or more antibiotic courses had a risk more than seven times greater than that of those who received none.

These numbers are sobering and are consistent with other data, such as a Canadian study that showed double the risk of asthma in children who received antibiotics in the first year of life. When was the last time you heard a doctor tell you that taking antibiotics could increase your child's risk of developing IBD or asthma? The answer is never. But at a recent conference on the microbiome, one of the physician participants suggested that maybe we should require black-box warnings—the ones in bold print on the slip of paper included with prescription drugs that warn about serious risks—for all antibiotics.

I discussed our studies of asthma earlier. Other conditions linked with asthma include hay fever and eczema, or atopic dermatitis. Hay fever involves sensitivity to environmental allergens, such as pollens, cat dander, and roses. Also called allergic rhinitis, it leads to sneezing and sinus problems. Eczema can be seen as patches of red skin or dry, scaly skin. Although several sites are prominent in children—scalp, face, and chest—it may be present everywhere.

The prevalence of both hay fever and eczema has been rising dramatically in recent years, paralleling the increase in asthma. In fact, many children start with eczema and end up with asthma (the so-called march to asthma), or they have all three conditions simultaneously. Millions of children in the United States alone are affected by these modern plagues. As I discussed earlier, mounting evidence suggests that lack of *H. pylori* is contributing to the rise in these conditions in childhood, but there could be other similarly missing microbes as well.

■ ■ ■

Other diseases also appear to have increased. Allergies to nuts used to be a very rare condition, but now as many as one in fifty children has the condition. Between 1997 and 2008 the percentage of children with diagnosed peanut allergy more than tripled. Although most cases are mild, and overdiagnosis is likely, allergic reactions can be severe, sometimes leading to sudden death. Even a trace of peanut in a food can be sufficient to set off a severe attack. That's why manufactured foods today carry a label announcing "produced in a facility in which nuts are present" or "certified nut-free." Nut allergies are changing the lives of tens of thousands of children worldwide. Where are these allergies coming from? I don't think that it is lack of pets.

Recently I've begun to think about a possible suspect. Recall that virtually all antibacterial agents exert similar effects promoting growth in farm animals; the animals get bigger whether they receive penicillins, tetracyclines, or macrolides. This led me to believe that all antibiotics produce more or less equivalent harmful collateral effects on our human resident bacteria. However, what if all antibiotics are not created equal? In fact, why should they be? We know that all major classes of antibiotics differ in terms of which bacteria they are good at killing. In our mouse experiments, we consistently find that we get stronger effects from tylosin, a macrolide, than from the penicillin drugs. We chose to study these two classes of drugs—the macrolides and the penicillins (the original members of a class of antibiotics called beta-lactams because of their chemical structures)—because together they represent more than 80 percent of all antibiotics prescribed for our children.

In the past, the macrolide of choice for kids was erythromycin. But compared to amoxicillin, a type of penicillin, it was less effective against some important pathogens and often led to side effects such as nausea and upset stomach. In 1991 two new macrolides were licensed in the United States: clarithromycin and azithromycin. Both were superior to erythromycin in all important ways and soon replaced it. Azithromycin is long-acting; even a few pills can have an effect for a week. Its manufacturer understood its great value and created the Z-pak, a small collection of drugs that can be dispensed as a single prescription and used for the entire antibiotic course. It is simple, effective, and has an easy-to-remember name.

In 1990, the year before it was licensed, the use of the Z-pak was essentially zero. By 2010, the last year for which I was able to find records, its use had climbed to nearly 60 million courses, and azithromycin now is the number one–selling antibiotic in the United States, even supplanting the pink liquid of amoxicillin. What that means is that every year about one in five of us in the United States is taking a course of azithromycin.

In 2010 more than 10 million courses of azithromycin were prescribed to children under eighteen and nearly 2 million just for those under the age of two.

For a drug that wasn't even invented twenty-five years ago, it has gained a remarkable foothold in the medical community. Lots of our epidemics seem to have gotten worse in this time period. Could these newer highly effective macrolides be playing a role? For me, it is just a hunch, but the mouse data are consistent, and the Centers for Disease

Control maps of antibiotic use in the United States also point a finger at macrolides. Interestingly, macrolide use is highest in the states with the highest obesity.

And then there is autism, a disease that strikes terror into the hearts of parents as its incidence continues to soar. When the disorder was first described by Dr. Leo Kanner in 1943, it was uncommon. Today, about one in eighty-eight children has autism or autism spectrum disorder (ASD). Although overdiagnosis is likely contributing to the rise in cases, it is not enough to explain the enormous increase. When differences in diagnostic criteria are taken into account, the disorder has still grown fourfold from the 1960s to the present.

Autism falls on a spectrum, from high-functioning children to those who are severely impaired. Basically, autistic brains are wired differently. Many complex interactions, especially those involved in communicating with other people, in understanding nuances and non-verbal cues, are impaired. Young children need these abilities to learn how to understand social contexts, and these skills get more and more important as they grow to adolescence and adulthood.

As with our other modern plagues, multiple theories abound to explain the increase in autism cases, including toxins in food, water, and air; exposures to chemicals and pesticides during pregnancy; and particular characteristics of the fathers. But no one knows. That there are so many theories indicates what a mystery it is.

My theory rests on the fact that gut microbes are involved in early brain development.

Your gut contains more than 100 million neurons—around the same order of magnitude as the number of brain cells—that operate more or less independently from the brain in your skull. They are found in two weblike layers between the muscles of your digestive tract, where they contract to propel and mix intestinal contents. They help keep things moving. The signals go right to your brain, but these nerve cells also can sense what's going on in your gut, most simply whether it is bloated. A rich network of nerve endings in your intestinal wall sends signals directly to your brain via the vagus nerve.

Another recent group of research papers, again using models in rodents, shows that this signaling from bottom (gut) up (to the brain) can affect cognitive development and mood.

These neurons, a part of the enteric nervous system, have regular contact with the microbes in your gut. There is an enormous amount of cross talk. One of the more interesting aspects of these brain-gut interactions is that your gut contains cells that make the neurotransmitter serotonin, which is involved, among other things, in the regulation of learning, mood, and sleep. We think of serotonin as being made and trafficked in the brain, but the neuroendocrine cells in your gut actually make 80 percent of it. And the bacteria in the gut are talking to these neuroendocrine cells, either directly or through inflammatory cells they recruit. In any event, it is an active conversation. Many of the microbes in the gut also make chemicals that the developing brain needs to function normally. These include gangliosides, small carbohydrate-like molecules that nerve cells use to build their coats.

Now think about what happens when a child takes antibiotics. If the composition of the microbes that lead to the production of gangliosides and serotonin is perturbed, the brain will be perturbed. There might still be a conversation among the microbes, gut wall, and brain, but it may be in the wrong language. In an adult, this might not make a big difference, but in a newborn baby or young child whose brain is developing rapidly? Although we do not know which is cause and which is effect, extensive studies point to abnormal serotonin levels in the blood of autistic children.

We know that antibiotics affect the development of metabolism (think of obesity) and immunity (as with asthma or Type I diabetes), so it's not a stretch to think they affect the complex development of the brain as well. This is a critical area for study, and we have begun our own lab work on the problem.

The final connection I want to make between changes to the microbiome and modern plagues is at this point still mostly theoretical. Antibiotics affect hormones, estrogens in particular. This was first noted when oral contraceptive pills were developed in the late 1950s. Women on the pill who received an antibiotic to treat an infection sometimes developed breakthrough bleeding; their periods might start in the middle of the month. It was quickly found that their estrogen levels had dropped. How could antibiotics do that? You guessed it: microbes were involved.

When estrogen is produced in the body—in both women and men but much more in women—it enters the bloodstream and is carried to the liver. There the estrogen is conjugated, meaning that liver cells add another compound, often a sugar, to the estrogen molecule. Then the conjugated estrogen is excreted by the liver into the bile, and from the bile it goes out into the intestine. If nothing interferes with its passage through the intestine, this so-called excess estrogen is excreted from the body as a component of feces.

On the other hand, the conjugated estrogen passing through the intestine might meet bacteria that see it as food. Such bacteria could easily cleave off the conjugate to nourish themselves, spitting out naked estrogen. This form of estrogen is readily reabsorbed by intestinal cells and ends up circulating back to the liver. So the fate of that estrogen molecule in the intestine depends on whether it meets a microbe that uses it for a meal or not. The presence of the microbe is the "switch" that determines whether the estrogen goes out of the body or is reabsorbed.

Thus, the composition of our gut microbes and their metabolic capacity have important bearing on our estrogen status. Dr. Claudia Plottel and I have called the microbes that affect estrogen "the estrobolome." An important question is whether the estrobolome of today is the same as it has always been or whether it has changed in recent years as a result of antibiotics and the like. While the answer is not yet in, we do know that girls are reaching the age of their first period, or menarche, much earlier than they used to. Moreover, the breasts of young women are noticeably bigger than they were in the past, more women are having problems with fertility, and the rate of breast cancer is increasing. For each of these issues, multiple factors could be playing a causative role but a change in overall estrogen metabolism or in the proportions of the subsets of estrogens (we have at least fifteen of these) that are excreted and absorbed could be an important factor.

As for breast cancer, two decades ago researchers identified mutations in two human genes, BRCA1 and BRCA2, that markedly increase breast cancer risk. Women who possess one or the other of these genes have an extremely high (greater than 50 percent) likelihood of eventually developing the disease. But BRCA-positive women born after 1940 tend to develop breast cancer at a much earlier age than BRCA-positive women born before 1940. Something in the environment has shifted, and it's not their genes. At this point, the role of an altered estrobolome is just speculative, but we are paying attention to it in the lab.

To reiterate my central idea, as our resident microbes succeed each other, we develop with them as an integrated circuit that includes our metabolism, immunity, and cognition. But we face unprecedented insults to our resident microbes. Although it may seem that I am blaming antibiotics and other modern medical practices for everything, including the kitchen sink, in fact I am only pointing to the diseases that have risen quite dramatically in the late twentieth century, the period in which these practices have been deployed. And indeed they all might have separate causes—and likely do—but there may be a single factor providing the fuel for each, tipping many people from a clinically silent stage into overt illness. It is like the loss of a reserve when the bank account of defenses is so depleted that with any new expense the account now is overdrawn. I believe that factor is the change in the composition of our microbiome, our resident organisms, just at the time when children are developing. And as we speculated five years ago, the changes in one generation may be compounding into the next.

To make matters even worse, I believe we may be heading toward a situation that I call antibiotic winter. This is an analogy to Rachel Carson's brilliant *Silent Spring*, where she predicts that birds could go extinct due to pesticides. But we could be traveling down a similar path.

15.

ANTIBIOTIC WINTER

Peggy Lillis, a fifty-six-year-old Brooklyn native, worked many jobs, sometimes two at a time, while raising two sons. For her last few years, she was a kindergarten teacher, the kind you always remember with deep affection. In late March 2010, Peggy had a minor dental procedure; by mid-April she was dead.

Peggy's dentist had prescribed a weeklong course of the antibiotic clindamycin, which is often given to ward off dental infections. Toward the end of the week, she developed diarrhea. Working with little kids, Peggy thought that she had the "stomach flu" and stayed home from work. But the diarrhea continued for four more days. Her family encouraged her to keep up with her liquids, and she contacted her doctor over the weekend. He arranged for her to see a gastroenterologist on Tuesday. But when Tuesday arrived, Peggy was too weak to climb out of bed, and her family called an ambulance. When the paramedics arrived at her home, they found her nearly in shock.

At the hospital, a colonoscopy revealed that Peggy had a severe infection involving the anaerobic bacterium *Clostridium difficile*. *C. diff*, as it is called, can be found in very low concentrations in the colons of healthy people. It usually minds its own business. But *C. diff* can wreak terrible damage when competing bacteria in the gut are wiped out by antibiotics. In a compromised colon, *C. diff* spreads like wild-fire. It can double itself every twelve minutes and dominate the intestine in a matter of hours. *C. diff* produces two or three toxins that it uses to coax the epithelial cells lining the colon to do its bidding. This helps it live but injures human cells. When the toxins spew forth, the colon becomes as porous as toast.

No one knows where Peggy picked up her *C. diff* infection. It could have been her own *C. diff*, or she could have gotten it from some-one close to her. Many patients in the hospital acquire it from another patient or from the hands of a health-care worker, but Peggy was not in the hospital. If your colon is healthy, *C. diff* should be blocked by the normal bacteria in your gut.

The antibiotic that Peggy took wiped out many of her normal bacteria. *C. diff* flourished and weakened her bowel wall. Fecal con-tents seeped through the wall of her bowel into areas that usually are bacteria-free. She became septic and spiked a high fever. Ironically, her treatment involved more antibiotics to clear her sepsis. When that wasn't sufficient, her doctors, in desperation, took her to the operating room to remove most of her injured colon. Despite heroic attempts, Peggy died in the hospital, less than a week after falling ill and less than two weeks after her dental procedure. How could this active, healthy, vibrant woman be gone so quickly?

We have known about antibiotic-associated diarrhea for more than fifty years, though *C. diff* was discovered to be the major cause only in the late 1970s. Most cases occur in people who are hospital-ized. This makes sense since they are often intensely exposed to antibiotics. Moreover, *C. diff* spreads by forming spores that can land on any surface or loft through the air. Thus hospitals, full of sick people, can be highly contaminated with *C. diff*. Tests show that hospi-tals often have a single strain circulating; other times, many different strains are present. Regardless, a single course of the correct antibiotic is sufficient to quell the infection in many patients.

But a single antibiotic course is not enough for up to a third of the afflicted; they relapse. And after retreatment, they may relapse again. This can happen thirty times and is sometimes so debilitating that patients waste away and die. Thankfully there is a new solution to the problem of relapse, which I will describe shortly.

It's not difficult to understand why relapses occur so often. As long as a person's intestinal ecosystem remains disrupted by antibiotics, chances are that these fast-multiplying organisms will bloom again. That the best treatment is more antibiotics only increases the disturbance. It's almost surprising that there is no relapse in two-thirds of C. *diff* patients.

Through the 1990s, with better infection-control practices in hospitals, such as more hand washing by health-care workers, better mopping of the floors, and isolating patients with severe diarrhea, rates of C. *diff* infections declined. But the problem could not be eradicated.

Over the past decade, patients admitted to our hospitals are on average sicker than they have been in the past. Chemotherapies are more successful, but there are more side effects. Patients are surviving more complicated surgeries, but recovery takes longer. Transplantation saves lives but requires immunosuppressive drugs, making people vulnerable to infection. The result is that more hospitalized patients receive more drugs of all kinds, including agents that suppress gastric acid and gut motility, and, of course, more antibiotics, often multiple kinds, simultaneously and sequentially.

A recent study of nearly 2 million hospitalized adult patients examined the use of the fifty most common antibacterial drugs prescribed. Across all of the patients studied, the investigators found that there were 776 days of therapy for every 1,000 patient-days in the hospital. These numbers include people coming in for normal procedures, like scheduled medication courses and blood transfusions, for which antibiotics are usually not used. This enormous load of antibiotics had to have had some kind of effect on our collective microbiome, and indeed it has.

And about ten years ago *C. diff* infections became more severe; more people died. What was going on? Analyses show that the strains had changed. A small segment of DNA just upstream from the toxin gene had been deleted. As a result, those strains spewed out more toxin, with all of their damaging effects.

Even more remarkable to me is that several different strains have different deletions, but all of them lead to greater toxin production. To a biologist, this means that extremely strong pressures are operating on *C. diff*, selecting hypertoxigenic strains over normal toxin producers. That several clones have mutated in parallel at the same time points to some common change in the environment. These same highly toxic clones are present in Europe and in North America, suggesting that hospital environments common to developed countries could be a factor. Indeed, hospitals are dangerous places.

What we did not foresee is how quickly *C. diff* infections could spread into the community. People like Peggy Lillis, who was not hospitalized, are becoming ill, and some are dying. Like a lion escaped from the zoo, *C. diff* has escaped the confines of the hospital and is now loose in the community. And the same clones, as passengers in someone's body aboard a jet plane, have crossed the oceans and set up shop in new communities—no passport required. In the United States, at least 250,000 people are hospitalized each year for *C. diff* infections that they acquired there or at home, and 14,000 die as a result.

The same thing has happened with MRSA, the infection due to antibiotic-resistant *Staph* that felled the two football players whose stories I told earlier. Twenty years ago, MRSA was found almost exclusively in hospitals, causing infections like the one suffered by the professional football player after his knee surgery. But now people with no exposure to hospitals, like the young high school player, are becoming infected. More virulent MRSA strains are appearing. That the two crises—*C. diff* and MRSA—have such similar characteristics and have arisen at more or less the same time tells us that our human microbial ecology is undergoing dramatic changes.

These are chilling stories, but sadly they are a harbinger of worse to come. The spread of these pathogens outside their "natural" reservoir, the hospital, to the larger community and across oceans represents a grave threat to our health. Finding ways to stop the spread of these lethal microbes must be a top priority.

The Centers for Disease Control and Prevention issued a landmark report in September 2013 that gave the first overall picture of drug-resistant bacteria in the United States. It ranked eighteen microbes according to their threat level and named three as "urgent." At the top of the list is a relatively new group of microbes called CRE, short for carbapenem-resistant enterobacteriaceae, which kill a high number of those infected and are resistant to essentially every antibiotic thrown at them. Moreover, CRE have the ability to spread resistance genes to other microbes by having microbial "sex" with them. CRE already have been identified in health facilities in forty-four states. *C. diff* and drug-resistant gonorrhea came in second and third on the list. MRSA was ranked as "serious," with eighty thousand infections a year and eleven thousand deaths.

Dr. Tom Frieden, who heads the center, warned that "antimicrobial resistance is happening in every community, in every health care facility, and in medical practices throughout the country. At least 2 million people per year in the U.S. get infections that are resistant to antibiotics, and 23,000 die. This is what happens," he said, "when microbes outsmart our best antibiotics." Saying that we face "catastrophic consequences" from overusing antibiotics, he added, "The medicine cabinet may be empty for patients with life-threatening infections in coming months and years."

■ ■ ■

We have a cabin in the Rocky Mountains. It sits on a ridge within a wide valley ringed by tall peaks. These are high mountains with snowcapped summits nine months a year and patches of ice still present in the depths of the summer. The trees make the mountains green until at the highest altitudes they peter out, and the summits are barren. It is a timeless, rugged, and magnificent landscape.

Until recently, the forests have been thick—too thick—replete with trees of all ages: majestic pines reaching straight to the sky like two-hundred-foot arrow shafts surrounded by fir, blue spruce, and groves of aspens. Everywhere you could see the baby trees coming up on their flanks, their branches almost tender with bright green, soft needles.

But about ten years ago, a pine beetle invaded our valley. It was probably always there but was held in check by intensely cold winters. Now, as the climate has warmed, the beetle has come back with a vengeance and is eating its way through the forest, ravaging entire mountainsides. Ninety percent of the trees are dead, waiting for a fire to turn them to ash.

What is happening to this Colorado landscape is a compelling metaphor for my missing-microbe hypothesis. Like the pine beetle, human pathogens surround us all the time, but their spread depends on certain conditions. How easily can they be transmitted from individual to individual? What is the host density and how susceptible are the hosts to attack? How healthy is the community? And what happens when the ecology changes, not in a forest, but inside a person? What happens when humans lose biodiversity? And what if that loss includes "keystone" species that keep ecosystems stable?

In the early 1950s, decades before *C. diff* was identified as causing antibiotic-associated diarrhea, Marjorie Bohnhoff and C. Phillip Miller conducted a series of experiments to determine the role of the *normal flora*—the term back then for our resident microbes—in fending off disease-causing bacteria. They believed it would be protective, and they tested their hypothesis by feeding mice *Salmonella enteriditis,* a species of *Salmonella* that causes disease in both mice and humans. When they gave the strain to normal mice, it took about one hundred thousand organisms to infect half of them. But if they first gave the mice a single oral dose of an antibiotic—streptomycin—and then several days later gave them *Salmonella*, it took only about three organisms to infect them. This isn't a 10 or 20 percent difference; it's a thirty thousand–fold difference. Welcome to the world of bacteria.

Miller and colleagues continued the work, showing that the effect wasn't limited to streptomycin. Other antibiotics, including penicillin, did the same. Even if the last antibiotic dose had been weeks earlier, the animals still could be infected by fewer microbes. In the sixty years since these experiments, many other researchers have confirmed and extended the findings. At least in mice, exposure to any of a number of different antibiotics increases susceptibility to an infection that sometimes is lethal. But is the same phenomenon true for humans?

In 1985 there was a massive outbreak of *Salmonella* infections in Chicago. At least 160,000 people became ill and several died. What could cause an event that would affect so many in one locality? Usually there are two main culprits, water or milk. Chicago had a municipal water system that was tightly regulated and policed; it was not the likely suspect. Besides, some of the people who fell sick were never in the city; they lived in suburbs that had their own water systems.

Thus suspicions turned to milk, which careful investigation confirmed. In particular, drinking milk from one grocery chain, the ubiquitous "Supermarket A," was implicated. Within days it became clear that milk from that chain was the source of the outbreak and that all of the milk was from their single large dairy. This dairy, a massive industrial facility with miles of pipes and huge vats, which I visited and inspected as an expert for the class of victims that brought a lawsuit, produced more than 1 million gallons of milk a week.

Most germane to our story is that the health department studied a group of fifty victims of the outbreak (cases) and fifty unaffected people (controls). They asked this simple question: Have you received any antibiotics in the month prior to becoming ill? They found that people who had taken antibiotics at any time during the month prior to the outbreak were five and a half times more likely to become ill than people who drank the milk but had not recently received any antibiotics.

Just as Bohnhoff and Miller showed in mice decades earlier, antibiotic exposure left people more susceptible to becoming ill from *Salmonella*. In the PAT experiments I described a few chapters ago, the mice were given their last doses of antibiotics on their fortieth day of life. But more than one hundred days later, we could still find evidence that their intestinal microbes were substantially perturbed.

It's not likely that Chicagoans were warned by their doctors that taking antibiotics would increase their susceptibility to infections, specifically to *Salmonella*. Has any health-care professional ever told you that? But increased susceptibility to new infections is one of the hidden costs of antibiotic use.

We now are in a good position to address one of the main questions raised throughout this book: How do antibiotics exert long-term effects on our resident microbes? In an earlier era, we relied on "indicator" organisms to represent overall microbial populations. An indicator organism is one that is used to estimate the presence of other microbial populations. For example, *E. coli* in surface water is an indicator organism for broader fecal contamination.

In 2001, my colleague in Sweden and good friend Dr. Lars Engstrand invited me to join in a study of how antibiotics affect indicator bacteria found in the human gut and on human skin. We used common colonizing bacteria that are easy to grow in culture: *Enterococcus fecalis* for the GI tract and *Staphylococcus epidermidis* for the skin. We asked whether people receiving a macrolide antibiotic—in this case, clarithromycin—as part of a one-week regimen to eradicate *H. pylori* from the stomach would exhibit an increase in macrolide-resistant bacteria elsewhere inside and on their bodies.

Unfortunately, the experiment worked beautifully. Before the subjects received the antibiotic, they had very low numbers of macrolide-resistant *Enterococcus* and *Staph epidermidis*, as did the control subjects, who were not treated at all. Things were different for the study subjects who received the antibiotic. Immediately following their treatment, the numbers of macrolide-resistant indicator organisms increased dramatically both in their feces and on their skin, but no such changes occurred in the untreated control subjects.

But our main question was how long would these blooms of antibiotic-resistant organisms last without any further macrolide exposure. The results were sobering. In the treated subjects but not in the controls, we found resistant *E. fecalis* three years later and resistant *S. epidermidis* four years later, which is when the respective studies ended, so we do not know how much longer the organisms would have persisted. I find it remarkable that a one-week course of an antibiotic can lead to persistence of resistant organisms more than three years later and in sites far away from the intended target of the antibiotic.

We also wanted to know whether the strains present at the beginning of the study were the same ones identified three years later, or if they had been replaced by new strains of the same species. Using DNA fingerprinting techniques, we found that in the beginning of the study each of the control subjects had a few different *Enterococcus* strains, which were mostly present three years later. However, in the treated group, the strains present before the treatment largely disappeared and were replaced by others. And over the course of the three-year study, strains with new fingerprints kept appearing. In other words, not only had we selected for resistance (which persisted), but we destabilized prior *Enterococcus* populations. We don't know whether those new strains had been present the entire time as minor populations or whether they were newly acquired but, in any case, the week of antibiotic treatment had a long-term and totally unintended effect on the stability of the particular strains of our indicator organism.

From the type of study we conducted, we cannot tell whether such changes lead to illness. If there is an effect, I would predict that the risk in most people would be small under usual circumstances. But we don't know the cumulative effect of many billions of doses of antibiotics given to hundreds of millions of people. Widespread treatments certainly enhance the pool of resistance genes, including those that can jump from our friendly bacteria to newly acquired pathogens. But the *Salmonella* experiments in mice, the Chicago outbreak, and the current epidemic of *C. diff* infections show us that antibiotic pretreatment increases susceptibility to pathogens. This is another hidden cost of changing our internal ecosystem.

■ ■ ■

It should be clear by now that even short-term antibiotic treatments can lead to long-term shifts in the microbes colonizing our bodies. A full recovery or bounce-back of healthy bacteria is in no way guaranteed, despite the long-held belief that such was the case. But that is not my only worry. I also fear that some of our residential organisms— what I think of as contingency species—may disappear altogether.

Recent research shows that people carry a small number of highly abundant species and a large number of much less common ones. For example, you may carry trillions of *Bacteroides* in your colon and only a thousand cells or fewer of other species. We are not sure how many rare, or contingency, species any of us has, but if you had only fifty or sixty cells of a particular type, it would be very difficult to detect them against the background of trillions of other bacteria.

The situation reminds me of *Where's Waldo?*, a children's book illustrating scores of people busily at work or at play with one character, Waldo, hiding in the crowd. The child's task is to find Waldo. If Waldo was a rare microbe and disappeared, we might not notice at all unless we looked for him specifically. When you take a broad-spectrum antibiotic, which is the kind most commonly prescribed, it may be the case that rare microbes occasionally get wiped out entirely. The critical point is that once the population hits zero, there is no bouncing back. As far as your body is concerned, that species is now extinct.

Why might it matter? By all rights, those puny species should be inconsequential. But microbes employ a powerful stratagem for their survival. Any small population of, say, a few hundred cells can explode into 10 billion or more cells by next week. The trigger for their massive bloom could be some component of a food you've eaten for the first time that only they have the enzymes to digest. Favored by a new, exclusive food supply, the rare microbe goes into overdrive and multiplies by 1 million percent. This blossoming could be good for you, too, because some of the energy captured in that new food by these microbes might end up in your bloodstream. But when food is in short supply, which was generally the case for most humans until quite recently, and people need to eat unfamiliar plants or animals, it would be useful to have a repertoire of enzymes to help us metabolize a wide variety of food chemicals. The genes of our flexible partners, our resident microbes, provide those enzymes.

Now consider what the consequences might be if one of your rare microbes went extinct. Imagine that it is ancient and has been dwelling in *Homo sapiens* for 200,000 years. One possibility is that it doesn't matter. Perhaps that microbe was a marginal player, so good riddance. Another possibility is that it is a "contingency" organism. You carry it and others in your baggage, not for everyday use but because it is like that set of hiking boots with crampons: useful when you need to walk across a glacier but dead weight the rest of the time. Or maybe it is that wedding dress from Grandma, beautiful but used only every other generation. Loss of such contingency species might not have much consequence except when you're suddenly faced with glaciers or weddings.

Another possibility is that you need contingency species only at certain times in your life, like that cane you keep in the attic for when you get old. In a way, the loss of contingency microbes represents the loss of biodiversity. Let's say that every cornfield in Iowa had the same high-producing strain of corn. For a while, all would be well; crop yields would be high. But if a pathogen emerged—a corn blight—that targeted that high-producing strain, then all of Iowa's cornfields would be susceptible to disease. In weeks, we could go from bountiful fields to acre after acre of diseased plants, and famine would not be far behind. Even small declines in biodiversity can make a community much more susceptible to an introduced pathogen. And as shown by the pine beetle and *C. diff*, it is the way of nature that pathogens are always present, with more just over the horizon.

An epidemic arising in one locale puts the whole world at risk.

We have seen this with the spread of influenza. When a new strain was recognized in Mexico in 2009, people in California and Texas fell ill days later, and others in New York City a few days after that. A few weeks later, this particular flu had spread across the United States and the rest of the world. We were relatively lucky that it was not a highly lethal strain, considering the hundreds of millions of people it infected. Its lack of potency notwithstanding, thousands of people all over the world still died. Even when a strain is not that virulent, when billions of people are infected the number of deaths add up. And if the strain is worse, as it was in 1918–19, then deaths are in the millions. We also were fortunate with the SARS epidemic in 2002, which was caused by a virus newly introduced to humans by animals, probably bats. Luckily its spread from person to person was not very efficient. It was devastating in a few places, but it died out from its own inability to transmit efficiently among humans. We dodged that bullet.

Our increased vulnerability to pathogens because of the "smaller" world in which we live is occurring just when our ancient microbial defenses are degrading. Such a conjunction provides fuel for raging conflagrations, either relatively local, like *Salmonella* or *E. coli* outbreaks, or potentially more global. The consequences of such a scenario are scarcely imaginable, but we have precedents to inform us. In the fourteenth century, the Black Death decimated Europe. We don't fully understand its cause, but changes in rodent populations were part of it. Another major factor was overcrowded, filthy medieval towns and cities, which were like tinder for the rat-borne plague. It lasted four years, and when it was over about one-third of the population of Europe had died, some 25 million people.

Another more recent plague, AIDS, has affected more than 100 million people worldwide since it came to us from chimpanzees. As terrible as HIV is, the disease does not pass easily from person to person like an influenza virus, so in one way—its speed of transmission—it is less frightening than a fast-spreading pestilence.

I am less interested in history books than in what might happen next. Plagues are inevitable wherever people congregate. This means that with a global population of 7 billion that is rising by 80 million a year—about the population of Germany—the questions are what will cause the next big plague, who will be affected, and when will it happen. Public-health measures will try mightily to minimize the costs, but it's possible we will be overwhelmed. The Great Influenza of 1918–19 killed tens of millions in an era without commercial airplanes and other forms of rapid mass travel to spread it. With a huge world population that is essentially contiguous, and with so many of us with weakened defenses because of our compromised internal ecosystems, we are vulnerable as never before.

I see many parallels between our changing climate and our changing resident microbes. The modern epidemics—asthma and allergic reactions, obesity, and metabolic disorders—are not only diseases but also external signs of internal change. We may see the problem when a child with an altered microbial ecosystem and diminished immune status encounters a mild pathogen that can easily damage the child's pancreas and lead to juvenile diabetes. Or the problem can show up when another child encounters a peanut or gluten, which are relatively recent additions to the human diet. Changes in their resident microbes and immune maturation similarly conspire to put them at risk, in this case for severe allergies to nuts or gluten. Like the worsening hurricane seasons we are seeing, these outcomes are bad enough, but they also are indicators of our larger imbalances, the loss of our reserves.

It's likely that a potentially deadly mutated microbe is now living in some animal somewhere in the world. It may have gained a new gene that helps it spread. Maybe it will crash into one of our domesticated animal species living in densely packed barns not far from where we live. Perhaps it will jump to an intermediary host, or maybe that new host will be us. In any case, the gathering storm is already here.

Fortunately we humans are on high ground when it comes to such storms; our diverse microbes with their 2 million genes in each of us help in resisting disease. They are the guerrilla warriors defending the home domain as long as we protect them. However, recent studies suggest that some otherwise normal individuals have lost 15–40 percent of their microbial diversity and the genes that accompany it.

This is the greatest danger before us: pathogens causing an epidemic against which we are helpless. Ecological theory tells us that the people whose resident bacteria have been disrupted the most will be the most vulnerable. All things being equal, the asthmatic, the obese, and those with the modern epidemic diseases will be the ones at highest risk to succumb. Human history and prehistory are pockmarked by prior plagues, and the world was more disconnected then. Genetic studies suggest that we come from a tiny founding population; our ancestors may have been survivors of some earlier cataclysm that was possibly related to climate change. But as contentious as the climate-change debate is these days, global warming may not be our biggest worry.

Unless we change our ways, we are facing an "antibiotic winter," a much greater peril, a worldwide plague that we cannot stop. Population biology is against us; we are no longer protected by isolation. We now live in one hugely connected village, and there are billions of us. And today many millions of us live with degraded defenses. When the plague comes, it could be fast and intense. Without high ground, like a river that overflows its natural banks, there is no sanctuary. All of this risk has been magnified by what in retrospect we will see as our reckless and profligate ways of abusing antibiotics. As concerned as I am about health problems such as diabetes or obesity, the most important reason why I am sounding an alarm is my fear of a catastrophic "antibiotic winter."

We talk about a pre-antibiotic era and an antibiotic era; if we're not careful, we'll soon be in a post-antibiotic era. This now is a major focus of the CDC, and I share its concern. But I am thinking about a different concept, not only the failure of antibiotics because of resistance but also the increased susceptibility of millions because of a degraded ecosystem. The two go hand in hand, but in a smaller interconnected world the second is a deluge waiting to happen and growing each day.

16.

SOLUTIONS

Last summer a relative called to ask me about a spreading rash on her leg. She e-mailed a picture: an ugly raised red blotch about two inches in diameter with a little dark spot in the center and a rim that looked like it had been marked by a highlighter. It resembled a bull's-eye. Because it was summer and she had been staying in Connecticut, two words came instantly to mind: *Lyme disease*.

I recommended that she start antibiotics immediately. She took her medication every day until the rash was gone and for several days more, even after she felt better, to complete the full course.

As hoped, the drug cured her infection, making both of us happy that we have effective antibiotics. I want to keep it that way. I'm not against antibiotics any more than I'm against ice cream—both great at serving their purpose—but sometimes there can be too much of a good thing. Overprescription of antibiotics and overreliance on C-sections are problems that urgently need resolution.

Solutions vary from the personal—the attitudes we embrace and the decisions we each can make—to the institutional—the kinds of policies the medical establishment or government should make and what kinds of research need to be prioritized. Sometimes the distinctions between personal and institutional blur, as in the case of antibiotics.

First we must curb our appetites for these powerful drugs. This is the biggest, simplest, and most achievable step we can take in the short term. It won't turn back the clock, but it could help slow the daily carnage to our microbial diversity.

Each of us can take personal responsibility for how to deal with antibiotics. Tell your doctor that you want to wait a few more days before taking amoxicillin for the cough that has lasted a week. Or you want to wait another day before you get a prescription for your child with a head cold. Resist pushing your physician for a quick fix to mitigate your anxiety. Without parental pressure, your doctor can make a better judgment about the need for an antibiotic.

Inform your dentist that you don't want to take antibiotics unless he or she can convince you that the benefits outweigh the potential risks. The axiom of good medicine (and dentistry) is "do no harm." Because we were not properly calculating the harm from antibiotics, they were above scrutiny. Many dental diseases are best managed by operative intervention and oral hygiene measures.

Stop using so many sanitizers on yourself and on your kids. While the key ingredient in these products, triclosan, is not an antibiotic, it kills bacteria on contact. What's wrong with good old soap and water? I use sanitizers only when I'm in the hospital seeing patients and during flu season. Most of the bacteria on my skin have been living with me for years. I know them, and they know me. I might pick up bacteria from other people, say on the pole in the subway car. Of course, I don't put my fingers in my mouth after touching that pole, but I don't use sanitizers either. I worry that I'll remove good bacteria, the ones that help me fight the bad bacteria that come my way.

Back to the question of what to do when your child is sick. I'm not saying wait and see in every instance. Sometimes children are quite ill and should be examined immediately. They are fussy, run high fevers, and gasp for breath. Or they are listless and don't respond normally to light or sounds. Their bellies may be swollen. They may have severe diarrhea or a terrible rash. These are true emergencies.

At a time like this, parents should carefully reconstruct the daily events leading up to the onset of symptoms and tell the physician everything they recall. After the exam, which may include blood tests and X-rays, many acutely ill children will need antibiotics immediately to avoid permanent injury or to save their lives. It would be a terrible mistake for a doctor to delay such treatment out of concern for causing collateral damage to resident microbes. Serious bacterial infections will always be with us.

Doctors thus face a conundrum: antibiotics are vital and yet way too many—more than 41 million courses in 2010—are being prescribed annually to U.S. children. Most kids don't need them.

Pediatricians and other health-care providers must be trained to think twice before prescribing antibiotics. They need to carefully weigh each situation. Is this a dangerous infection or, more likely, is it a mild one that will go away by itself in the vast majority of children?

This judgment call is not a simple task. It can take years of experience to conduct truly careful examinations. For the doctor in a hurry, it's far easier to write a prescription for everyone who walks in the door with a runny nose, sore throat, or red ear drums. It's more time-consuming for a doctor to examine a child carefully, to discuss with a parent why the antibiotic should be withheld, to answer questions, to explain the danger signals, and to say "call me in the morning if things don't get better."

Along with better training, pediatricians need better pay. Paradoxically, physicians involved in the primary care of our children— the doctors on the front lines who receive tens of thousands of visits every day from parents and their children—are among the lowest paid of all physicians in the United States. Something is wrong with our system when the doctor who performs a brief diagnostic procedure—some form of X-ray, for example, or a fifteen-minute operation—is paid many times more than the doctors making crucial decisions about our children's health.

Pediatricians should be paid sufficiently to methodically evaluate the children brought to them and be rewarded for taking time to discuss each diagnosis with mothers and fathers. Since our current system markedly undervalues this kind of care, it's no surprise that 70 percent of the kids coming in with what is considered an upper respiratory infection walk out with an antibiotic.

Many well-informed parents and good doctors and nurses are trying to change these attitudes and practices, but our system conspires against them. Unconscious biases are everywhere. We think that cutting routine office visits to twenty minutes, fifteen minutes, even ten minutes will save money when in fact, with less time for doctors to examine and less time to think, we are incurring far greater costs through excessive testing and needless treatment.

Physicians and patients also should be made aware of how local customs drive their prescription rates. Southerners swallow about 50 percent more antibiotic courses than people living in western states. I doubt there is a 50 percent difference in the incidence of bacterial diseases in those regions. Like rates for C-sections or episiotomies, such differences in usage reflect variations in the practice of medicine.

■ ■ ■

When I talk with colleagues about what has to happen to change attitudes about our health practices, they are downright pessimistic. Time horizons, they say, are distressingly long. Habits are engrained. People are terrified of germs. Doctors like to feel powerful and at the same time are afraid of being sued. Government regulators fear making difficult decisions that might invite political controversy or jeopardize their careers. Health systems are paid by insurers and by the government to act, not to withhold treatment. And pharmaceutical companies are content with a status quo that provides them hefty returns for little or no new investments.

But I am hopeful that change will come faster than the naysayers think. I believe that we are at a tipping point. As discussed in the previous chapter, the director of the Centers for Disease Control recently called a press conference to focus on resistance to antibiotics; magazines are full of stories about horrible cases of antibiotic-resistant infections; and many people are beginning to realize that "germophobia" has serious downsides. As we consider real costs and limited benefits, simple actions become more sensible.

Governments can do more to bring antibiotic use under control. The French provide a shining example. In 2001, France had the highest rate of antibiotic use among European countries, which prompted its public-health agencies to swing into action. In 2002, the French National Health Insurance launched "the national plan to preserve antibiotic efficacy," which was solely aimed at preventing the spread of antibiotic-resistant organisms.

Of course, to reduce resistance they had to reduce use. While hospital patients received many antibiotics, more than 80 percent of the drugs were being prescribed to people in the community. That was the place to intervene. The major target: reduce the use of antibiotics given to children for viral infections of the respiratory tract. Health officials focused on the winter months, when most such infections occur.

Called Antibiotics Are Not Automatic, the campaign was aimed at simultaneously changing the mind-set of both patients and their health-care providers. Because France has a centralized database of pharmacy prescriptions, health officials were able to review a large sample, roughly between 2002 and 2006, during which 453 million antibiotic prescriptions—nearly 10 million per month—were written. For a country of 60 million people, that is a lot of antibiotics.

By the end of the intervention in 2006–2007, the prescription rate had declined by 26 percent. This reduction, which was repeated across all areas of France, affected nearly every class of antibiotics and worked across all ages, not just children as targeted. But it was particularly effective in kids under three whose annual prescription rates fell from about 2.5 courses per child to about 1.6, a drop of 36 percent.

Other public-health authorities in France have gone further. A pilot campaign in the French Alps called Antibiotics Only When Necessary is a logical next step from "not automatic" to "only when necessary." If the United States adopted a similar program, we might also slowly wean ourselves of our addiction. Already antibiotic use in children is off from its peak, perhaps by 20 percent, based on programs originally intended to decrease antibiotic resistance. The programs mostly involve the education of doctors and other health practitioners on the front lines about why they should avoid the reflexive urge to pull out their prescription pads. And in Sweden, a highly developed medically sophisticated country, the outpatient prescription rate is "only" 388 per thousand persons versus our 833. That antibiotic prescribing rates in Sweden are less than half of ours shows that it can be done without excessive health hazard.

■ ■ ■

There's one more thing the government can do to reduce overuse of antibiotics: prevent farmers from giving them to animals whose products—meat, milk, cheese, eggs—we eat. The carryover of antibiotics into our food and water is completely avoidable. We must set a date, which can be a series of dates with increasingly stringent requirements, to ban the practice.

For consumers, it means that the prices of meat, eggs, milk, and fish will all go up at the supermarket cash register by a small percentage. In comparison, we are already paying the price of antibiotics in food through the spread of resistant organisms and the diminishing utility of our antibiotics, and we are probably contributing to our costly epidemics of allergy and autoimmunity and metabolic problems. In the future, we can pay at the supermarket or we can pay at the clinic through our insurance premiums, taxes, and compromised personal health.

In late 2013, the Food and Drug Administration announced that it will take the first steps to removing growth-promoting antibiotics from our livestock. The change is based on the threat of antibiotic-resistant bacteria traveling from animals to humans, but a collateral benefit would be to reduce antibiotic residues in our food and drinking water. Although this is an important move in the right direction, we must hold the FDA's (and the industry's) feet to the fire, because without enforcement the producers could use similar amounts of antibiotics to "treat illness" in the livestock.

And let's not stop at antibiotics. Food producers are allowed to sell foods that have detectable levels of antiworm agents, insecticides, and hormones. Interestingly, for certain hormones like testosterone and estrogen, there are no set limits, because of the following wording in the World Health Organization regulations: "Residues resulting from the use of this substance as a growth promoter in accordance with good animal husbandry practice are unlikely to pose a hazard to human health." Is this now an appropriate standard for us?

■ ■ ■

The way new antibiotics are developed also needs overhauling. For inspiration, we can go back about a century, when Paul Ehrlich, the early germ-theory pioneer, experimented with hundreds of compounds until he found Salvarsan, a safer derivative of arsenic, which was his "magic bullet" for treating syphilis. It was good for only that, nothing else. When you get a skin abscess, you may have been exposed to many bacteria, but almost always just a single microbe dominates the infection. If your therapy were directed narrowly at that one bug, you'd get better.

But for more than seventy years, pharmaceutical firms have sought "broad-spectrum" agents that kill many types of microbes. There are plenty of advantages to this approach. When someone is sick—with pneumonia, a urinary infection, or an infected wound—the doctor can immediately start treatment with a drug that kills all of the expected bad actors. And if occasionally one medication won't cover everything, then a second, and rarely a third, can be added. This works most of the time. But the broader the antibiotics and the more they are used, the bigger the collateral effects on our resident bacteria.

There are two problems with narrow-spectrum drugs. First, very few exist. We need to create and test them. If we want an antibiotic that is specific for *Streptococcus pneumoniae*, we have to identify a target in that organism that is shared by few if any other bacteria. Same for *Staph aureus*.

Second, even if we had one antibiotic for each of the thirty or forty species that cause most bacterial infections in humans, we wouldn't know which one to use in any individual case. The coughing patient does not come in with a sign that reads "I am infected with *Streptococcus pneumoniae*." Right now our diagnostic tests are slow, taking days or longer. Doctors need rapid tests, enabling them to sample blood, sputum, exhaled air, or urine to look for the chemical signature of particular organisms. With that information, your doctor could reach into a formulary and take out the best narrow-spectrum agent for your condition.

The good news is that it should be relatively simple to develop narrow-spectrum agents. We may have to target just one organism at a time and experiment with chemicals or even bacteriophages (viruses that eat bacteria). Bacteriophages that can be produced by the trillions do the same work as antibiotics and have been battling (and living with) bacteria for billions of years. I'm currently advising a company that is developing a type of medicine similar to phages because I believe it will lead to a new armory of narrow agents.

We also can tap into more than a decade of genomics. We have deciphered the genetic sequence of all of the major human bacterial pathogens. We know which genes are found in each organism and the potential structure of the products they make as if we had a map guiding us to the buried treasure. We can look for genes that are unique to *S. pneumoniae*. We might find specific inhibitors for particular enzymes and be able to create a "designer" antibiotic.

The bad news is that these new drugs will be expensive. For manufacturers to recoup their outlays, each five- to ten-day course of narrow-range antibiotics used in relatively few people would have to be priced at thousands of dollars, compared with the tens of dollars today for broad-spectrum drugs. Given our current economic model, this is not feasible. The pharmaceutical industry is focused on developing drugs that millions of people take every day for years—like pills for high blood pressure, diabetes, heart disease and its prevention— or ultraexpensive drugs for patients with cancer.

On the diagnostic side, there has been significant improvement of late. Currently under development are new panels of diagnostic tests that can distinguish between viral and bacterial infections much better by identifying specific agents. And a new class of diagnostics is being brought to market that use the host's immune responses as the indicator of which organism is causing the trouble. Both of these are in their early stages, but the path to their widespread use is clear. The only issue is money.

But in the long run it may be more expensive to ignore the need for better diagnostics and narrow-spectrum antibiotics. If taking antibiotics early in life is leading to some portion of the cases of obesity, juvenile diabetes, asthma, and other disorders, what is the lifetime cost of those illnesses in dollars, not to mention suffering and years of life lost?

We can pay now to prevent or we can pay later to treat. The drugs and diagnostics that I propose would be public goods, with value for practically everyone well into the future. It's a little like road building. Let's say we needed a highway between Los Angeles and Phoenix. No one can afford to construct it alone, but if we build it collectively through taxes we have Interstate 10. The quality of life has improved a lot for those who live there and perhaps a little for the rest of us who might one day like to travel at high speed across the desert. Similarly, we need a national or international initiative to create the diagnostics and the therapies we need. We live in a highly interconnected world. I was shocked to learn that antibiotic use in China is even higher than it is in the United States.

C-sections are another overused medical practice that can benefit from personal and institutional changes. If you are a woman of child-bearing age, carefully question the need for elective C-sections. Is it best for your baby? Ask your doctor if it is absolutely necessary. Obviously, if your doctor tells you that you need an emergency C-section to save your baby's life or your own life, then don't hesitate.

Recently, I spoke with a friend whose daughter was about to give birth. She knew my stance. "And remember, no C-section . . ." I said at the end of our conversation.

"Only if absolutely necessary," she agreed. "If she does have a C-section, either she or I will use the gauze-in-the-vagina technique and inoculate the baby ourselves."

The gauze-in-the-vagina technique is a practice that my wife Gloria is studying in Puerto Rico. The idea is simple. Because a baby born via C-section misses picking up microbes from the mother's vagina, the deed can be done artificially. The mom or a helper places a gauze pad in the vagina so it collects bacteria-laden secretions and then, right after birth, gently swabs the baby's skin and mouth. It's not exactly the same as a vaginal birth, but microbiologically it's a step in the right direction.

I believe that Gloria's technique or some variants of it will become the standard practice within a few years. This is not to say that it's perfect or won't cause new problems. A few babies might get infections from their mothers. These might have happened in any case, but there will always be suspicion that the swabbing technique caused them. We must screen mothers for potential pathogens, and if we begin to routinely swab newborns delivered by C-section, then we'll need to monitor outcomes across all of the relevant time frames, including the long term. Maybe one day we will understand which of the mom's organisms are crucial and just give those to the baby, but I doubt it. In my opinion, most of them, in their diversity and plurality, may be useful.

Meanwhile, health providers are slowly starting to wake up to the need for change. I predict that doctors will be more cautious about advocating for C-sections as they learn more about the consequences. With increasing data, hospitals and insurance companies will be more reluctant to accept the high C-section rates. One day, parents of a child who has developed a problem attributed to an elective C-section— maybe obesity or juvenile diabetes or autism—will sue the doctor and hospital for malpractice. That will really get people's attention. Currently the fear of being sued is for *not* doing something: not getting an X-ray, not prescribing an antibiotic, not doing a C-section. Soon there will be the fear of getting sued because of unnecessary and unjustified actions. Fear is one of the great equalizers.

■ ■ ■

As I travel the country talking about missing microbes, many people ask me what I think about probiotics. Are they what they're cracked up to be? When should people take them and for which conditions?

A few years ago, a colleague of mine—a healthy woman in her midsixties—woke up doubled over with pain in her lower abdomen. She had a fever and worried that she might need an operation. But after blood tests and X-rays, she was diagnosed with diverticulitis, an inflammation of the lower part of her bowel. This is a relatively common condition, especially in older people, but no one really knows what causes it. Often requiring hospitalization, it usually goes away by withholding food, resting the bowel, and taking a course of antibiotics.

Why the antibiotics? Because they work. As such, the conventional explanation is that by suppressing the overall gut "flora" or particular but unspecified bad actors, the inflammation subsides. That probably is correct, but the details are still missing.

In the case of my colleague, the terrible pain came back five times in separate episodes. She feared that something awful was happening inside her. After the fifth episode, she consulted a gastroenterologist who suggested she take a probiotic. She takes it every day and has had no episodes in the past two years.

Coincidence? Maybe, maybe not. When she told me her story

much later, I was glad to hear that a probiotic worked in her case. Presumably, the cultures changed some sort of microbial equilibrium in her intestines. But the fact is we can't explain their mechanisms of action, if any, because we cannot directly see the interior dynamics of the human gut.

Despite my colleague's success story, I'm generally skeptical about the many claims surrounding all the probiotics crowded on our grocery store shelves, pharmacies, and health-food stores. They are almost completely untested. In our free country, it turns out that marketing probiotics is a kind of freedom of speech. The packages make all sorts of vague claims about health promotion, yet in most cases no rigorous trials were done to show that the ingredients were actually effective.

The definition of *probiotics* is broad, but so are the different types of bacterial cultures sold in stores. Sometimes they consist of single strains of bacteria; other times they are mixtures. They may be sold as liquids, powders, or salves. Sometimes what are apparently the same strains are sold under different labels with different text extolling their benefits. Some of the cultures were originally isolated from milk and milk products. Others, like *Bifidobacter*, originate from human babies, and still others from human adults. Combinations abound. It's the Wild West; the field is almost completely unregulated.

The best that I can say is that they are generally safe: you can take them just as you would food, and if you are a normal, generally healthy person, the risk of a problem is small. But do they work? Many people swear by them, so on some level some of them must, but I cannot tell you which ones.

And then there are prebiotics. Unlike living probiotics, prebiotics are chemical compounds that stimulate the growth of organisms we consider favorable. For example, as discussed earlier, human milk is naturally full of prebiotics, including small sugar molecules that can only be used by particular bacteria present in the GI tract in babies. By their very presence in human breast milk, they select for the growth of the initial founding bacteria that colonize the early gut.

Chemists have used these and related formulations as prebiotics to stimulate bacteria that people already have in their gut.

Synbiotics are mixtures of probiotics and prebiotics. The prebiotic increases the chances that the probiotic will colonize the intestines in greater numbers and for a longer time. ·

The theory behind probiotics, prebiotics, and synbiotics is appealing, but the current ways they are used smacks of placebo effects. Doctors used to give sugar pills or injections of salt water or shots of vitamin B_{12} (to people who had normal B_{12} levels), and, believing they were getting real medicines, the patients would feel better. Placebos are notoriously effective. They work for many people, particularly in conditions where attitudes play a role, such as lower-back pain. Pain, which can be devastating also, sometimes may be no more than opinion on the state of the body.

Some products claim they will make you feel better, perkier, more energetic. But this goal is vague, hard to define, and even harder to test. How do you know that you are feeling better and, most important, compared to what?

When you go to a health-food store looking for probiotics, the very act of walking in suggests you are seeking something to make you feel better. By buying the product, you are ready to be helped, and the placebo effect kicks in.

We won't know if these products are doing any more good than placebos until we conduct blinded clinical trials. Subjects would be given a probiotic or a placebo that looks, smells, and tastes identical without knowing which was which. The study would look at the health effects, if any, of each treatment. Unfortunately few rigorous trials of this nature have been carried out. Manufacturers who make good money selling probiotics are disinclined to pay for such studies.

Another assertion is that a probiotic will help with a specific disease, say ulcerative colitis or cancer, or speed recovery from influenza. These claims, by their very nature, are easier to test. But few of the well-conducted trials that have been performed show efficacy.

It's not hard to see why. Certain diseases like ulcerative colitis, for example, have a variable course in individuals and among patients. A study would require a large number of patients, maybe one hundred or more, to tease out the variation and to see any substantial effect. And that would be expensive.

I'm not dismissing probiotics. In fact, I think they will be very important in future disease prevention and treatment, but we need to have a much stronger scientific base for their efficacy. Which organisms exactly should we put back into our bodies? Maybe your lost bacteria are different from mine. How do we know what is suppressed in you or in danger of extinction? Since antibiotics suppress or eliminate some microbes, I predict that in coming years we'll give people probiotics as a form of standard care to accompany the course of needed antibiotics. But first we must understand which microbes we're dealing with.

■ ■ ■

Remember the tragic case of Peggy Lillis, the healthy woman who died from a *C. diff* infection? This awful problem continues unabated, but recently a new technique has shown remarkable success in treating people with multiple relapses of the disease.

Called fecal microbiota transplantation (FMT), it is the deliberate transfer of feces from one person to another. Of course, the very thought of this procedure is revolting, but it has saved lives, especially for people with recurrent *C. diff* infections.

To administer this treatment, the doctor obtains a fecal specimen—a fresh bowel movement from a healthy person who might be a relative of the patient or someone who is just a "good" donor whose poop has already helped many people. The doctor makes a slurry of it in a salt solution and then gives the resulting opaque brown liquid to the patient with the *C. diff* infection. It is delivered via a plastic tube or by endoscope through the nose down into the stomach or duodenum or in the other direction via colonoscopy or as an enema in the rectum.

Although the practice conjures up disgusting images, it works. A number of doctors have been doing this for some years, and in 2013 a pivotal and attention-getting study from the Netherlands was published in the widely read *New England Journal of Medicine*. The investigators conducted a randomized clinical trial in patients with recurrent *C. diff* infection; the participants were offered the chance to be treated either conventionally with antibiotics or with a fecal transfer. The cure rate for those getting the drugs was 31 percent, whereas for those who chose the fecal transfer it was 94 percent. The difference was so substantial that the trial was stopped because it would have been unethical to give the remaining subjects the standard treatment.

This well-conducted, rigorous trial established a "proof of principle" that restoring microbes to people with a damaged intestinal ecosystem—as occurs with *C. diff*—could be good medicine. With this evidence, investigators now can conduct trials to find the active "ingredient," that is, which microbe or group of microbes are necessary to reverse the illness. The near universality of success using many different "donors" indicates that the key ingredients are common to all of us. It might be a single group of organisms or a variable group in which substitutions are permitted, like the Chinese restaurant where you can pick one from column A and one from column B.

The other great benefit of the Dutch study and the work that went before it by pioneers like Drs. Alexander Khoruts and Lawrence Brandt is that it establishes fecal transfer and its future variations as potential therapies for other illnesses in which a deranged intestinal ecosystem, such as inflammatory bowel disease, celiac disease, and irritable bowel syndrome, may play a role. And it's not far-fetched to think that it also could be used to treat obesity and a variety of immunologic disorders, possibly including autism. If a disordered intestinal microbial community is at the root of these problems, then restoration via fecal transfer could be a solution.

After the Dutch study, many desperate people began "do-it-yourself" fecal transfers at home, by enema. We don't know if anyone has been harmed or how many people have been helped. In

2013 doctors providing the procedure were cautioned by the FDA to adhere to a series of regulatory hurdles designed to ensure safety. I believe that ruling was quite reasonable. The history of medicine is scarred by many episodes of the overly enthusiastic embrace of something that seemed so good, like DES or thalidomide, only to lead to real harm. This is especially important when considering the transfer of biologic materials from one person to another. The transfer of AIDS and hepatitis via blood and blood products reminds us of the peril. However, if we could give pure cultures of probiotic bacteria, the problems of human-to-human transmission would be obviated.

■ ■ ■

Now consider that most of our children may be growing up without the full complement of their necessary microbes. Where can we find the right ones to put back? Perhaps models of the assembly of the microbiota in developing mice can teach us the key principles. Are there places in the world where people have not been exposed to antibiotics? If so, perhaps their gut bacteria would be fully intact. Maybe we could turn them into medicine. Maybe the excrement of our fellow world citizens, people who have had the fewest exposures to antibiotics, antiseptics, and the accoutrements of modern life, the ones who live deep in the Amazon or in the highlands of New Guinea, are the ones we need for the microbe transfers. Are their microbes different from ours?

Gloria found answers in Venezuela. In 2008 an army helicopter pilot spotted a tiny village in the endless High Orinoco jungle. It didn't appear on any map. The helicopter descended with an interpreter on board who spoke their native language. He told them that they were friends and that the government wanted to bring medicines. They said that they had seen helicopters in the sky before, and from other members of their tribe in other villages they had heard the word *medicina*. But they had never before seen people who were not of their tribe.

When the team surveyed the little village, they found two metal objects: a machete and a can. The people had traded for these objects with other Indians and had learned about the power of the *medicinas*. They wanted medicine because they had their share of misfortunes.

The villagers' contact with the outside world was inevitable, and the Venezuelan government made the good decision, in my opinion, to vaccinate them. Measles and flu would eventually come to the village and be lethal. So under the auspices of multiple permits and with ethical committee approvals, a medical team made arrangements to return later. Gloria requested that they ask the villagers for specimens that she could study. When they returned to the village, with doctors and health workers ready to vaccinate and treat infections, they also obtained mouth and forearm skin swabs from thirty-five villagers across all ages as well as fecal specimens from twelve of them. Through close cooperation with the Venezuelan authorities and with investigators and institutions in Amazonas State with whom Gloria has worked for more than twenty years, those swabs were sent to her lab for careful investigation.

What a treasure those specimens were. Gloria now had resident microbes from people who were essentially from the Stone Age, with no written language, no mathematics, no contact with the modern world. They had never taken antibiotics. In a sense, their microbes were living fossils. The fecal samples were absolutely unique—and priceless.

A few years later, DNA from the specimens had been extracted and sequenced. One morning in our dining room in New York, Gloria and her fellow researchers Rob Knight and José Clemente pored over the latest analysis of the fecal samples. In their three accents—Spanish, Kiwi (New Zealand), and Venezuelan—they boisterously discussed the colorful plots of microbial populations in the intestines of the 12 Amerindians in comparison with those of 157 representative young adults and their families from Colorado. The graphs appeared one after another on José's computer.

The differences were stark and almost paradoxical. The 157

North Americans had only a few taxa that were unique to them, while the 12 Amerindians had more than a hundred unique species that were not present in most of the U.S. subjects. Plus they had more taxa than the U.S. subjects by far, even though many species were found in low numbers. How to explain this asymmetry? One interpretation is that many of the microbes they carried had disappeared from us as a result of our exposures to antibiotics and other aspects of medical care and, indeed, of modern life.

Here, too, was important evidence supporting my nearly twenty-year-old hypothesis. The graphs were visually beautiful and the contrasts clear. It did not take complex statistical analyses to see the major differences between the samples from the two populations. Someday these ancient microbes, missing from us, might be used to protect our children from the modern diseases now plaguing them and that the villagers are not suffering from. One day we might give them back to our babies to fill the modern void.

■ ■ ■

As with fecal transfer, the idea is to somehow restore the missing microbes. They might come from faraway places or from your own family. I can imagine grandmothers who have not taken so many antibiotics in their lifetime giving their bacteria to their grandbabies.

I imagine that babies in the future might undergo a new kind of workup. For the one-month checkup, the doctor will examine the baby along with its poop and urine. In the lab, the bacteria will be sequenced and counted, and the urine analyzed for particular metabolites. Out will come a statement: baby doing fine but need to supplement with *Bifidobacterium*. And for another baby it might be *Allobaculum*, and for the third *Oxalobacter*. So the doctor will order up from the formulary the optimal culture for that child. And for another child, the composition of the cultures likely will differ.

Maybe these microbes will be applied to the mom's nipple so that the baby can take them in with mom's milk. Or maybe they will get a special dosage of formula, let's say with some cells of *Oxalobacter* and oxalate, the nutrient that this microbe loves but that we can't digest. Such a "synbiotic" approach will help get the strain started: the probiotic with its prebiotic. These are not just random organisms. In my lab at NYU, we are studying each of these and their relationship with humans.

In 1998 I predicted in the *British Medical Journal* that we would one day be giving *H. pylori*, a disappeared organism, back to our children. Since then, the support for this idea has only grown deeper and the list of disappeared organisms longer. But these are early days in the discovery process; most of the workings of the mechanisms are still secret.

EPILOGUE

Karl Benz, Henry Ford, and the other automotive inventors in the late nineteenth and early twentieth centuries made a monumental contribution to human life. They invented, perfected, and mass-produced the internal combustion engine, a machine that enables us to drive to work, to carry large loads, to go on holidays, to explore the world, and much more. Human existence has changed as a result: we are more interconnected, we can war at longer range, we can meet people of all different ethnicities and cultures.

We already know that the internal combustion engine also has spawned a host of new problems or worsened those we already had: air pollution, vehicular homicide, traffic jams. Perhaps Ford could have anticipated these; although they were unintended, they could have been imagined. Cities with horse-drawn carriages had traffic tie-ups, and all of the excreta was not exactly pleasant. As such, many of the problems that followed the widespread introduction of the internal combustion engine were extensions of the known.

Yet imagine that about a hundred years ago someone told Henry

Ford that every time a person turned the ignition in his or her car, the ice cap in Greenland would melt a little. It would have been unfathomable, and Ford likely would have dismissed it immediately. What if someone told *you* that same thought about thirty years ago? You probably also would have thought that it was ridiculous. How could those two occurrences be related? Yet we know how the unconnected become connected. This is one example of how our successful inventions are transforming our "macroecology," the status of our planet.

The story I have told concerns how we are changing our "microecology" with well-intended and indeed life-saving measures like antibiotics and Cesarian sections. That the resident microbes living in us are changing with disastrous results may seem as foreign as global warming would have been to Ford. But now, more than forty years after the "Earth" movement began, I believe that we are finally primed to contemplate and address these changes.

The ill effects in this story may be no less profound than those related to global warming and, in fact, may be operating on a shorter time frame. I do not wish to ban antibiotics or Cesarian sections any more than anyone would suggest banning automobiles. I ask only that they be used more wisely and that antidotes to their worst side effects be developed. The truth is always obvious in retrospect. How could people really have thought that the sun revolves around Earth or that Earth is flat? Yet dogma are powerful and to their adherents infallible.

Once the question, Do antibiotics have a biological cost as well as clear benefits? is even posed, the horizon begins to shift. The answer is that, of course, our powerful antibiotics could affect our friendly bacteria. Of course, changing the mechanics of labor and delivery from the ancient ways to the modern in a third or half of our births today could have effects. Of course, purposely removing our natural microbial inhabitants is likely to have complex consequences.

The logic is inescapable. Our ancient microbes are there for a reason; that's how we evolved. Everything that changes them has a potential cost to us. We have changed them plenty. The costs are already here, but we are only just beginning to recognize them. They will escalate.

The moment for substantive change is now. But change takes time, and reversal of the losses takes even longer. As with global warming, there is the risk that the status quo is "locked in." Yet I am optimistic. The changes in human microecology have been going on for only about a century and especially the past sixty to seventy years. This is the blink of an eye in the totality of human experience. Change that comes fast can depart just as rapidly.

We stand at the proverbial crossroads. We have medicines and practices that have served us well but have had unintended consequences. With powerful agents of any kind, there always are unintended consequences, so it should be no surprise. But the wake-up call is that we are not talking about uncommon events. The practices that endanger our children are at the core of modern health care.

We have made so much real progress in combating and eradicating terrible diseases. But now perhaps our efforts have peaked, and the fruits of discovery have left their seeds, indigestible and toxic. We must act, for the consequences are beginning to swallow us, and stronger storms lie ahead.

Yet many types of solutions are available. And with some of these, there may be synergies, combining the effects of two approaches, like curtailing both C-sections and antibiotic use, and eventually replacing disappeared organisms. For the future of our children and theirs, it is time for us to begin implementing them in earnest.

NOTES

1. MODERN PLAGUES

1   **have been getting healthier:** In ancient times, one-third to one-half of children did not survive until the age of five. (See T. Volk and J. Atkinson, "Is child death the crucible of human evolution?" *Journal of Social, Evolutionary and Cultural Psychology* 2 [2008]: 247–60.) Childhood death rates remained high through the nineteenth century. Even by 1900, in some U.S. cities up to 30 percent of infants died without seeing their first birthday. (See R. A. Meckel, *Save the Babies: American Public Health Reform and the Prevention of Infant Mortality, 1850–1929* [Baltimore: Johns Hopkins University Press, 1990].) By the twentieth century, improved public health started to make a huge difference; infant mortality went from about 100/1,000 in 1915 to about 10 in 1995 (*Morbidity and Mortality Weekly Report* 48 [1999]: 849–58). Childhood mortality rates have continued to fall in the last half century (G. K. Singh and S. M. Yu, "U.S. childhood mortality, 1950 through 1993: trends and socioeconomic differentials," *American Journal of Public Health* 86 [1996]: 505–12).

3   **worldwide obesity epidemic:** Although increased body mass ultimately reflects more calories in than out, obesity is a complex issue. The question of whether all food calories are equal in terms of human metabolism is controversial. Issues such as physical and psychological stress and lack of sleep may affect *(increasing)* food intake. Lack of exercise may play a role in weight gain disproportionate to its direct effect on calorie expenditure. Maternal smoking, prenatal environment, hormone disruptors, and salted-food addiction all have been postulated as causative, and even chemical toxins have been considered to play a role. (P. F. Baillie-Hamilton, "Chemical toxins: a hypothesis to explain the global obesity epidemic," *Journal of Alternative and Complementary Medicine* 8 [2002]: 185–92.)

3   **has risen 550 percent since 1950:** In developed countries, juvenile (Type 1) diabetes has been steadily rising. (V. Harjutsalo et al., "Time trends in the incidence of type 1 diabetes in Finnish children: a cohort study," *Lancet* 371 [2008]: 1777–82.) Although, after more than fifty years of continued growth and a recent period of accelerated growth, the incidence appears to be leveling off, possibly because of public-health activities. (V. Harjutsalo et al., "Incidence of type 1 diabetes in Finland," *Journal*

of the American Medical Association, 310 [2013]: 427–28.) Worldwide, the annual increase in Type I diabetes in recent years has been about 3 percent. (P. Onkamo et al., "Worldwide increase in incidence of Type I diabetes—the analysis of the data on published incidence trends," *Diabetologia* 42 [1999]: 1395–403.)

6   **resemble those of adults:** T. Yatsunenko et al., "Human gut microbiome viewed across age and geography," *Nature* 486 (2012): 222–27. In this study, after comparing the gut microbiota from people in the United States, Malawi, and Venezuela (Amerindians), researchers found that the compositions in infants were markedly different from those in adults. But as children matured, their microbiomes became more and more adultlike. Importantly, the age at which this happens is three. The transition from no microbiota to an adultlike microbiota is all accomplished during the earliest stages of life, just as many functions in the host are developing.

6   **the "disappearing microbiota":** The disappearing-microbiota hypothesis evolved over a number of years. A few of my key papers that develop the theme include: "An endangered species in the stomach," *Scientific American* 292 (February 2005): 38–45; "Who are we? Indigenous microbes and the ecology of human disease," *EMBO Reports* 7 (2006): 956–60; with my very distinguished colleague Stanley Falkow, "What are the consequences of the disappearing microbiota?" *Nature Reviews Microbiology* 7 (2009): 887–94; "Stop killing our beneficial bacteria," *Nature* 476 (2011): 393–94.

8   **"cloak of invisibility":** The discovery of the stealth mechanisms of *Campylobacter fetus* involved a progressive series of experiments conducted over nearly twenty years. A few of the key papers include: M. J. Blaser et al., "Susceptibility of *Campylobacter* isolates to the bactericidal activity in human serum," *Journal of Infectious Diseases* 151 (1985): 227–35; M. J. Blaser et al., "Pathogenesis of *Campylobacter fetus* infections. Failure to bind C3b explains serum and phagocytosis resistance," *Journal of Clinical Investigation* 81 (1988): 1434–44; J. Dworkin and M. J. Blaser, "Generation of *Campylobacter fetus* S-layer protein diversity utilizes a single promoter on an invertible DNA segment," *Molecular Microbiology* 19 (1996): 1241–53; J. Dworkin and M. J. Blaser, "Nested DNA inversion as a paradigm of programmed gene rearrangement," *Proceedings of the National Academy of Sciences* 94 (1997): 985–90; Z. C. Tu et al., "Structure and genotypic plasticity of the *Campylobacter fetus sap* locus," *Molecular Microbiology* 48 (2003): 685–98.

9   **and house cats (*Felis catus*):** Unfortunately, taxonomy is often complicated because our house cats also have been classified as *Felis silvestris*, within the species of wildcats, or sometimes called *F. silvestris f. catus*. Still, a cat by any other name would meow.

9   **natural defenses against it:** Based on our studies of variation in campylobacters and host responses to them, we began to study the same for the gastric campylobacter-like organism (or GCLO), which for a time was called *Campylobacter pyloridis*, then *Campylobacter pylori*, before eventually its current name, *Helicobacter pylori*, was agreed upon. Our first papers about this were: G. I. Pérez-Pérez, and M. J. Blaser, "Conservation and diversity of *Campylobacter pyloridis* major antigens," *Infection and Immunity* 55

(1987): 1256–63; and G. I. Pérez-Pérez, B. M. Dworkin, J. E. Chodos, and M. J. Blaser, "*Campylobacter pylori* antibodies in humans," *Annals of Internal Medicine* 109 (1988): 11–17. From these studies we developed a blood test (which is the basis for most of the blood tests used today in the United States) to diagnose whether or not a person has *H. pylori* in his or her stomach.

9 **"the only good *H. pylori* is a dead *H. pylori*"**: In response to my paper in the *Lancet* (M. J. Blaser, "Not all *Helicobacter pylori* strains are created equal: should all be eliminated?" *Lancet* 349 [1997]: 1020–22), David Graham wrote to the editor: "The only good *Helicobacter pylori* is a dead *Helicobacter pylori*" (*Lancet* 350 [1997]: 70–71). This became the signature concept for the present era.

10 **our normal gut flora:** *Flora* is the old name for the countless organisms that live in humans. We used to call them our normal flora. But bacteria are not plants, and the organisms that live in and on us are both small and diverse. We now call these organisms our *microbiota*. And all of the relationships between the microbiota and ourself, and with each other, are collectively called the *microbiome*.

2. OUR MICROBIAL PLANET

13 **"his middle finger erases human history":** J. McPhee, *Basin and Range,* book 1 in *Annals of the Former World* (New York: Farrar, Straus & Giroux, 1998).

13 **a few exceptions that reinforce the rule:** H. N. Schulz et al., "Dense populations of a giant sulfur bacterium in Namibian shelf sediments," *Science* 284 (1999): 493–95. But such large microbes are the anomalies in a world dominated by microscopic forms.

15 **the distance between corn and us:** N. Pace, "A molecular view of microbial diversity and the biosphere," *Science* 276 (1997): 734–40. To Carl Woese, Norman Pace, and many others, bacteria were at the very origins of all life on Earth.

16 **240 billion African elephants:** W. B. Whitman et al., "Prokaryotes: The unseen majority," *Proceedings of the National Academy of Sciences* 95 (1998): 6578–83; J. S. Lipp et al., "Significant contribution of Archaea to extant biomass in marine subsurface sediments," *Nature* 454 (2008): 991–94; and M. L. Sogin et al., "Microbial diversity in the deep sea and the underexplored 'rare biosphere,'" *Proceedings of the National Academy of Sciences* 103 (2006): 12115–20.

17 **selection in action:** Plastic-eating bacteria. T. Suyama et al., "Phylogenetic affiliation of soil bacteria that degrade aliphatic polyesters available commercially as biodegradable plastics," *Applied and Environmental Microbiology* 64 (1998): 5008–11; E. R. Zettler et al., "Life in the 'plastisphere': microbial communities on plastic marine debris," *Environmental Science and Technology* 47 (2013): 7137–46.

18 **water, and bacteria—loads of them:** T. O. Stevens and J. P. McKinley, "Lithoautotrophic microbial ecosystems in deep basalt aquifers," *Science* 270 (1995): 450–54.

19 **the common intestinal bacterium *E. coli*:** *E. coli's* formal name is *Escherichia coli*, honoring Theodor Escherich, a German doctor who discovered it in 1885 in the feces of healthy people, and called it *Bacterium coli commune*. In the early twentieth century, the

name was changed to *Escherichia coli*. Although the best-known bacteria in the human gastrointestinal tract, it usually represents less than one-thousandth of all the bacterial cells present. While most *E. coli* strains are harmless, there are distinct strains that can cause several different types of disease. Because of the ease of growing *E. coli* in culture, it has become a model organism to study the biology, biochemistry, and genetics of cellular life. Many of the five thousand genes in *E. coli* cells have analogues in human cells.

21 **"and ever shall be, until the world ends"**: In 1993 S. J. Gould wrote a review that appeared in *Nature* about E. O. Wilson's then new book *The Diversity of Life*, in which he indicates that Wilson already knows that rather than an individual age of reptiles or of mammals, these are but parts of the eternal age of bacteria, as he so states. (S. J. Gould, "Prophet for the Earth: Review of E. O. Wilson's 'The diversity of life'," *Nature* 361 [1993]: 311–12.)

3. THE HUMAN MICROBIOME

23 **They are symbionts:** Symbiosis, defined in the nineteenth century, is the close relationship of two (or more) species living together, sometimes for most or all of their lifetimes. Although it may mean living together harmfully, neutrally, or helpfully, it also can be used to describe just the mutually helpful relationships. A species party to such a relationship is a symbiont.

23 **Aphids, small insects that live on plants:** N. Moran, "The evolution of aphid life cycles," *Annual Review of Entomology* 37 (1992): 321–48.

24 **more apelike than cowlike:** H. Ochman et al., "Evolutionary relationships of wild hominids recapitulated by gut microbial communities," *PLOS Biology* 8 (2010): e1000546.

25 **Of fifty known phyla:** A phylum is a term in biology referring to the taxonomic classification between kingdom and class. The kingdom Anamalia, encompassing all animals, has about thirty-five phyla, ranging from Arthropoda (insects) to Chordata (having a spinal cord, like humans).

25 **in your mother's womb, you had no bacteria:** This has been the long-held belief, but evidence is beginning to emerge that even in the womb in many animals microbes are normally present (L. J. Funkhouser and S. Bordenstein, "Mom knows best: the universality of maternal microbial transmission," *PLOS Biology* 11 [2013]: e1001631). However, this is still an area of controversy. We will probably know for sure, one way or the other, in humans in a couple of years.

25 **over the first three years of life:** In a study of the gut microbiota of healthy people in three places—the United States, Malawi, and Venezuela (Amerindians)—Yatsunenko and her colleagues, including my wife, Gloria, catalogued which microbes were present across people of all ages. In early life, there were great similarities between the three different ethnic groups, but as they got older they diverged. Perhaps most important, the composition of the microbiota in infants is very different

from that of adults, but gradually it becomes more and more adultlike, reaching adult levels by the age of three! (T. Yatsunenko et al., "Human gut microbiome viewed across age and geography," *Nature* 486 [2012]: 222–27.) Initially I was surprised, but the more I thought about it, the more sense it made—the microbiome develops in parallel with the development of the child. This was consistent with my hypotheses about the importance of the early-life microbiota.

26 **are home to different species:** We did the first survey of the skin using molecular methods beginning in 2004 and showed the incredible diversity but also the symmetry between left and right. (Z. Gao et al., "Molecular analysis of human forearm superficial skin bacterial biota," *Proceedings of the National Academy of Sciences* 104 [2007]: 2927–32.) Then, using more powerful methods, other investigators confirmed and extended the observations, showing more subtle differences between left and right hands, and how our computer keyboards carry the microbial signatures of our fingertips—that is, we can tell your keyboard from mine (N. Fierer et al., "Forensic identification using skin bacterial communities," *Proceedings of the National Academy of Sciences* 107 [2010]: 6477–81). They also showed that each of the three major types of skin—dry, moist, and oily—has its own major populations (E. A. Grice et al., "Topical and temporal diversity of the human skin microbiome," *Science* 324 [2009]: 1190–92), and that a single group of fungi dominate in most of our skin, except for the bottom of our feet (K. Findley et al., "Topographic diversity of fungal and bacterial communities in human skin," *Nature* 498 [2013]: 367–70).

27 **250 healthy young adults:** The large Human Microbiome Project sponsored by the National Institutes of Health made incredible progress in laying out the fundamentals of our microbial composition. In the important study of healthy young adults in the United States (actually in Houston and St. Louis), the outlines of the human microbiome were shown. (C. Huttenhower et al., "Structure, function and diversity of the healthy human microbiome," *Nature* 486 [2012]: 207–14.) In this paper, there were nearly as many authors (me included) as there were subjects, but it was a very complex "big science" national effort that paid large dividends—and that will keep paying as more and more scientists use the trove of information accumulated, from sampling at sixteen sites in men and women, plus three vaginal sites in the women. From that study, for example, we know much more about the populations in the mouth: how the top of the tongue, hard palate, and cheek are more similar to one another than to the gingival crevices.

27 **they don't like oxygen:** The microbial composition of the gingival crevice is vast: its density is similar to that in the colon, and the variety of bacteria is enormous (I. Kroes et al., "Bacterial diversity within the human subgingival crevice," *Proceedings of the National Academy of Sciences* 96 [1999]: 14547–52; and ibid.). That interface between tooth and gum is where periodontal disease occurs, and the hope is that by better understanding the microbial populations and their dynamics, we will be better able to prevent or treat this major cause of tooth loss.

28 **who is attractive to mosquitoes:** N. O. Verhulst et al., "Composition of human skin microbiota affects attractiveness to malaria mosquitoes," *PLOS ONE* 6 (2011): e28991.

28 **dozens of species living there:** Z. Pei et al., "Bacterial biota in the human distal esophagus," *Proceedings of the National Academy of Sciences* 101 (2004): 4250–55. Until we published our paper, no one thought that the esophagus had any residential bacteria, only transients traveling from the mouth and throat down.

29 **colonic bacteria and in their functions:** Just as we can construct family trees of plants and animals, using new computational tools, we can do the same for the bacterial populations living in different ecological niches. We can compare the composition of microbial populations living in freshwater ponds to those in the oceans. (Not surprisingly, they are quite different.) When such tools are applied to the compositions of the colonic microbes in, for example, mice and humans, we can see enormous parallels (R. E. Ley et al., "Worlds within worlds: evolution of the vertebrate gut microbiota," *Nature Reviews Microbiology* 6 [2008]: 776–88). At higher taxonomic levels, starting at the phylum, we are nearly identical, but as we descend the phylogenic ladder, the differences become greater until, at the species level, mouse and human are highly distinct. In a way, these microbial similarities and differences capture our evolution from a common ancestor to distinct species—*Mus musculus* and *Homo sapiens*—as well as our own genetic inheritance. Yes, even our cohabiting microbes inform us that "ontogeny recapitulates phylogeny," a concept in evolutionary biology I learned as a high school student, long before I could even guess what evolution was all about.

30 **the activities of your microbes:** W. R. Wikoff et al., "Metabolomics analysis reveals large effects of gut microflora on mammalian blood metabolites," *Proceedings of the National Academy of Sciences* 106 (2009): 3698–703. Investigators compared germ-free mice (born in bubbles and without any bacteria at all) and more conventionally raised mice. They used very sensitive chemical sampling and detection methods to examine the contents in the bloodstreams of the two groups of mice. Of nearly 4,200 chemical constituents, only 52, slightly more than 1 percent, were seen in the blood of the germ-free animals; the more than 4,000 others were ultimately derived from bacterial metabolism. These studies provided evidence that most of the chemical constituents in the blood of mice (and thus by extension in us) are ultimately derived from having microbiota and from their interaction with our cells.

30 **first chemical processing and then absorption:** H. J. Haiser et al., "Predicting and manipulating cardiac drug inactivation by the human gut bacterium *Eggerthella lenta*," *Science* 341 (2013): 295–98.

31 **at least for a short period:** R. Avallone et al., "Endogenous benzodiazepine-like compounds and diazepam binding inhibitor in serum of patients with liver cirrhosis with and without overt encephalopathy," *Gut* 42 (1998): 861–67.

31 **which is low in protein:** The sweet potato is about 2 percent protein, so an adult would have to eat about five pounds of sweet potato a day to ingest enough protein.

31 **essentially lacks *Lactobacillus*:** J. Ravel et al., "Vaginal microbiome of reproductive-age women," *Proceedings of the National Academy of Sciences* 108, suppl. 1 (2011): 4680–87.

33 **gut microbiome is relatively stable:** J. Faith et al., "The long-term stability of the human gut microbiota," *Science* 341 (2013): DOI: 10.1126/science.1237439. By studying the same people over time, often years, Jeff Gordon's lab showed that although there is turnover in the organisms that can be detected, there also is considerable stability. In their study, about 70 percent of the organisms present in adults on sampling were estimated to be present one year later.

34 **the changes in microbial populations were more significant:** Dr. Nanette Steinle from the University of Maryland presented the data on the dry bean/lentil study at the American Society for Nutrition's poster session on April 23, 2013. In other work, immediate diet effects on the microbiome were seen, but with long-term stability of overall composition. (See G. Wu et al., "Linking long-term dietary patterns with gut microbial enterotypes," *Science* 334 [2011]: 105–8.)

34 **only as long as the person was consuming the special diet:** L. A. David et al., "Diet rapidly and reproducibly alters the human gut microbiome," *Nature* (2013): DOI 10.1038/nature12820.

34 **millions of unique genes:** Just as the human microbiome has been the focus of "big science" in the United States, another large group coalesced in Europe, the Meta-Hit consortium. They have done important work that is both unique and complementary to the findings of the HMP. J. Qin et al. ("A human gut microbial gene catalogue established by metagenomic sequencing," *Nature* 464 [2010]: 59–65) showed the huge range in composition from person to person. M. Arumugan et al. ("Enterotypes of the human gut microbiome," *Nature* 473 [2011]: 174–80) postulated that humans could be divided into three major types based on the composition of their gut microbiome, perhaps analogous to human blood types. Whether the typing scheme will stand up over time and whether the types are relatively stable in an individual host remain to be determined.

35 **by comparison, has about 23,000 genes:** While this is technically correct, because of alternative transcriptional and translational start sites and alternative splicing, the number of different proteins that we can actually make may be one or two orders of magnitude higher. Thus, the protein-coding properties of us and our microbiome may be more similar than originally thought.

35 **bacterial genes in subjects' guts varied dramatically:** In a recent paper by the MetaHit group (E. Le Chatelier et al., "Richness of human gut microbiome correlates with metabolic markers," *Nature* [2013]: 500, 541–46), 292 subjects were studied in terms of their gut microbial gene counts and their metabolic status. The results clearly show the gene count for the two groups: high for about three-quarters of the subjects and low for the remaining quarter. On average the people in these two groups differ significantly in their metabolic status. Those in the low-gene-count group were much more likely to have the metabolic syndrome, a constellation of findings associated with obesity,

diabetes, hardening of the arteries, and high blood pressure. One question that could not be resolved by the study is which came first, low gut microbial gene count or the metabolic syndrome. But a companion paper showed that dietary interventions that improve metabolic status raise the gene count (A. Cotillard et al., "Dietary intervention impact on gut microbial gene richness," *Nature* 500 [2013]: 585–88).

37 **staggering ten million–fold:** Qin et al., "A human gut microbial gene," showed this.

37 **never before encountered:** See I. Cho and M. J. Blaser, "The human microbiome: at the interface of health and disease," *Nature Reviews Genetics* 13 (2012): 260–70, where we more fully discuss the concept of contingency organisms.

38 **cows and their rumen bacteria:** The rumen is the specialized, first stomach in ruminants like cows and sheep. It is a specialized compartment in which the microbes that are present ferment the ingested feed, allowing its energy to be digested by the host. The rumen is also a prime example of symbiosis, and the resident microbes include bacteria, fungi, protozoa, and viruses.

39 **if you played fair and square:** See M. J. Blaser and D. Kirschner, "The equilibria that allow bacterial persistence in human hosts," *Nature* 449 (2007): 843–49, for a fuller exposition of these ideas about equilibrium relationships between our microbes and us.

4. THE RISE OF PATHOGENS

41 **causing a form of encephalitis:** Encephalitis means inflammation of the brain. It is usually an acute infection caused by a virus or a bacterium, but may be due to other organisms, or may be noninfectious.

42 **eat their prey from within:** D. Quammen, *Spillover: Animal Infections and the Next Human Pandemic* (New York: W. W. Norton & Company, 2012).

43 **and killed fifty:** The outbreak came out of nowhere, and uncounted thousands were exposed to the contaminated sprouts. A medical description of the outbreak was published in U. Buchholz et al., "German outbreak of *Escherichia coli* O104:H4 associated with sprouts," *New England Journal of Medicine* 365 (2011): 1763–70; and a description of the characteristics of the strain in C. Frank et al., "Epidemic profile of Shiga-toxin-producing *Escherichia coli* O104:H4 outbreak in Germany," *New England Journal of Medicine* 365 (2011): 1771–80; and how it all happened in M. J. Blaser, "Deconstructing a lethal foodborne epidemic," *New England Journal of Medicine* 365 (2011): 1835–36.

47 **epidemic diseases began to take off:** W. McNeill, *Plagues and Peoples* (New York: Anchor, 1977).

47 **might infect from one-third to one-half of those exposed for the first time:** A nineteenth-century accounting of what happened when measles came to an isolated island was by Peter Panum in his classic "Observations Made During the Epidemic of Measles on the Faroe Islands in the Year 1846" (*Bibliothek for Laeger*, Copenhagen, 3R., I [1847]: 270–344). There also have been more recent observations, e.g., when a boat landed in Greenland in the 1940s with a crew member who had measles.

47 **18 deaths every hour:** World Health Organization data on measles and deaths, http://www.who.int/mediacentre/factsheets/fs286/en/. Measles, the relatively mild and nearly universal childhood disease in the developed world until an effective vaccine was introduced in the 1990s, shows a very different face in developing countries. There, in the setting of malnourishment, immunodeficiency, and concurrent infections, measles is a killer. Each year, more than one hundred thousand children die as a result of measles. It is a calamity and one that vaccine can prevent. But the problems in deploying the vaccine to all in need have been political, logistic, and economic.

47 **human population of 500,000 people:** Decades before it entered the mainstream, Francis Black was one of the first to think about island biogeography in terms of the spread of infectious diseases in humans. (See F. L. Black, "Measles endemicity in insular populations: critical community size and its evolutionary implication," *Journal of Theoretical Biology* 11 [1966]: 207–11.)

48 **the measles virus quickly spread from person to person:** Panum, "Observations Made During the Epidemic of Measles."

48 **grain bins and trash heaps:** M. J. Blaser, "Passover and plague," *Perspectives in Biology and Medicine* 41 (1998): 243–56.

48 **broke out in Kinshasa, Zaire:** Not only in the fourteenth century but also in this one, plague still visits cities when the conditions are ripe for it. In Africa and India there has been urban plague in recent years. See, for example, G. Butler et al., "Urban plague in Zaire," *Lancet* 343 (1994): 536; and details of a continued endemic focus: P. Boisier et al., "Epidemiologic features of four successive annual outbreaks of bubonic plague in Mahajanga, Madagascar," *Emerging Infectious Diseases* 8 (2002): 311–16.

48 **Twenty percent of children did not survive:** Several different techniques were used to measure mortality. Extensive work was done by Samuel H. Preston and Michael R. Haines on estimating childhood mortality. See the chapter "New Estimates of Child Mortality During the Late-Nineteenth Century" in their book *Fatal Years: Child Mortality in Late-Nineteenth Century America* (Princeton: Princeton University Press, 1991), 49–87.

5. THE WONDER DRUGS

54 **difficult-to-treat malignancy:** In several studies, working with colleagues in Hawaii, at the Mayo Clinic, and in Japan, we had shown that people carrying that bacterium in their stomachs were more likely to have or to later develop stomach cancer (A. Nomura et al., "*Helicobacter pylori* infection and gastric carcinoma among Japanese Americans in Hawaii," *New England Journal of Medicine* 325 [1991]: 1132–36; N. Talley et al., "Gastric adenocarcinoma and *Helicobacter pylori* infection," *Journal of the National Cancer Institute* 83 [1991]: 1734–39; M. J. Blaser et al., "*Helicobacter pylori* infection in Japanese patients with adenocarcinoma of the stomach," *International Journal of Cancer* 55 [1993]: 799–802). Other studies, conducted in England by David Forman and in California by Julie Parsonnet, showed very similar results. In a couple of years, we changed what people knew about the cause of stomach cancer, then and now the number-two cause of cancer

death in the world (after lung cancer). We now know that more than 80 percent of all stomach cancer cases can be attributed to *H. pylori* (see chapter 9).

57 **from white blood cells and from saliva:** Fleming discovered lysozyme, a constituent of innate immunity, in saliva. It is an enzyme that by breaking the chemical bonds that hold the cell walls of bacteria together effectively dissolves (lyses) bacterial cells. It was a major discovery (in retrospect) of one of the arms of our innate (inborn) immunity. We have evolved a variety of molecules, like lysozyme, that have antagonistic activity against whole classes of bacteria. They reduce contamination on our mucosal surfaces, our coastlines, and also help clear tissues of invading bacteria. But most important, Fleming's discovery of lysozyme prepared him to recognize the lytic activities possessed by the accidental mold that landed on his plates a few years later. (A. Fleming, "On a remarkable bacteriolytic element found in tissues and secretions," *Proceedings of the Royal Society, Series B* 93 [1922]: 306–17.)

57 **They had been inoculated with staph:** "Staph" is a nickname used by health professionals to refer to *Staphylococcus*. Usually this is in reference to *Staph aureus*, a major pathogen, rather than "*Staph epi*" (*S. epidermidis*), an important colonizer of the skin, with low virulence.

58 **used molds to treat infected wounds:** Gloria's grandmother, who lived in rural Spain in the early twentieth century, would use moldy bread to help infections heal; it was common knowledge among the peasants, but never explored for how it worked.

58 **After publishing his results:** A. Fleming, "On the antibacterial action of cultures of a penicillium, with special reference to their use in isolation of B. influenzae," *British Journal of Experimental Pathology* 10 (1929): 226–36.

59 **the first sulfonamide:** Sulfonamidochrysoidine, the red dye called Prontosil, was shown by Domagk in 1932 to protect mice from *Strep*. It had actually been discovered more than twenty years earlier, but its medical use was not tested then. In 1935, a French group found that Prontosil was a pro-drug, metabolized to sulfanilamide, the active agent.

59 **but not good enough:** Co-trimoxazole, the drug that was successfully used to treat my case of paratyphoid fever, was actually a derivative of those original sulfa drugs. But they worked much better in combination than did the early forms of the drugs in the 1930s and '40s.

60 **a way of growing the penicillium molds:** When I later visited the Pfizer plant and research center in Groton, Connecticut, in the mid-1990s, the air was redolent with the odor of molasses. Why that sweet characteristic smell? Oceangoing ships from the West Indies would come up the Thames River, and each would dock with its hold filled with molasses, which was used as the main source of food for giant vats of the *Penicillium* mold that itself manufactured the life-saving penicillin.

60 **made by one living form to fight against another:** Penicillin, the first antibiotic, was produced by a mold as an antibacterial substance. The sulfa drugs were made by chemical synthesis in a factory. They are not technically antibiotics, because they

are synthetic, but we use the term antibiotics generically to include both true anti-biotics and chemically synthesized agents (same for fluoroquinolones, like cipro).

6. THE OVERUSE OF ANTIBIOTICS

64 **5.5 million stoves:** The statistics about our prosperity and pent-up demand in 1945–1949 come from the Public Broadcasting System's show *The American Experience* and the episode "The Rise of American Consumerism."

66 **derives from these ultracontagious human viruses:** See T. M. Wassenaar and M. J. Blaser, "Contagion on the Internet," *Emerging Infectious Diseases* 8 (2002): 335–36, for a discussion of the parallels between infectious diseases and so-called computer viruses, the malware that is transmissible from person to person (or rather from computer to computer).

66 **after a couple of weeks:** For a natural history of cough in URIs, see: S. F. Dowell et al., "Appropriate use of antibiotics for URIs in children, Part II: Cough, pharyngitis and the common cold," *American Family Physician* 58 (1998): 1335–42.

69 **to ward off rheumatic fever:** The purpose of antibiotics for treatment of children with strep throat or what appears to be strep throat is complex. In S. T. Shulman et al., ("Clinical practice guideline for the diagnosis and management of Group A streptococcal pharyngitis: 2012. Update by the Infectious Diseases Society of America," *Clinical Infectious Diseases* 55 [2012]: e86–102), the IDSA guidelines committee makes important recommendations that I paraphrase: they note that testing for Group A strep (GAS) usually is not recommended for children or adults who have an acute sore throat with clinical features that strongly suggest a viral cause (e.g., cough, runny nose, hoarseness, and mouth ulcers). Diagnostic studies for GAS are usually not indicated for children under three years old because acute rheumatic fever is rare in children under three, and the incidence of strep throat is uncommon in this age group. Laboratory confirmation is essential in making a precise diagnosis because physicians often greatly overestimate the probability that GAS is the cause of sore throat. A test that is negative for GAS provides reassurance that the patient's sore throat likely has a viral cause. While treatment early in the course leads to a more rapid clinical cure in patients with acute GAS pharyngitis and decreases transmission of GAS to other children, the predominant rationale for treatment of this self-limited illness is the prevention of acute rheumatic fever and other complications. The committee states that efforts to identify GAS carriers are not ordinarily justified, nor do carriers generally require antimicrobial treatment because they are unlikely to spread strep throat to their close contacts and are at little or no risk for developing acute rheumatic fever.

70 **follow the safer course:** Recommendations by the American Academy of Pediatrics (AAP) regarding antibiotic use have been in place for a long time. (See S. F. Dowell et al., "Principles of judicious use of antimicrobial agents for pediatric upper respiratory tract infections," *Pediatrics* 101, suppl. I [1998]: 163–65.) An important update was

issued last year: A. S. Lieberthal et al., "The diagnosis and management of acute otitis media," *Pediatrics* 131 (2013): e964–99.

70 **64 patients with pneumonia:** W. S. Tillett et al., "The treatment of lobar pneumonia with penicillin," *Journal of Clinical Investigation* 4 (1945): 589–94.

71 **antibiotics to people in the United States:** L. Hicks et al., "US outpatient antibiotic prescribing, 2010," *New England Journal of Medicine* 368 (2013): 1461–62.

71 **in the United States and other developed countries:** For antibiotic use in other developed countries, for example, see M. Sharland, "The use of antibacterials in children," *Journal of Antimicrobial Chemotherapy* 60, suppl. 1 (2007): i15–i26.

76 **MRSA stands for methicillin-resistant *Staphylococcus aureus*:** MRSA infections were noted in the 1960s, almost immediately after antibiotics like methicillin were used to treat people with *Staph aureus* infections. But such drugs were mostly used in hospitalized patients, and MRSA strains were largely confined to the hospital. But in recent years, MRSA has been spreading in the community. These days, among serious staph infections in people coming to emergency rooms for treatment, about 80 percent are due to MRSA. (G. J. Moran et al., "Methicillin-resistant *S. aureus* infections among patients in the emergency department," *New England Journal of Medicine* 355 [2006]: 666–74.) This is a dramatic change from the past. Resistance in *Staph* is spreading so greatly that the natural divisions between hospital and community have been blurred. But in fact the dominant MRSA strains are different. There are two largely separate populations of MRSAs, each adapted to its own ecological niche but each driven and selected by the enormous antibiotic pressures in both hospital and community.

76 **"a tiny little thing that I can not see":** Brandon Noble, IDSA website: http://www .idsociety.org/Brandon_Noble/.

76 **NCAA Division III championships:** Ricky Lannetti, MRSA awareness website: http://www.mrsaawareness.com/mrsaawareness/Home.html.

7. THE MODERN FARMER

80 **an endless game of chemical warfare:** See extended discussion of the arms races in chapter 2.

80 **which have similar core structures:** See V. D'Costa et al., "Antibiotic resistance is ancient," *Nature* 477 (2011): 457–61; and K. Bhullar et al., "Antibiotic resistance is prevalent in an isolated cave microbiome," *PLOS ONE* 7 (2012): e34953.

80 **resistance from our activities:** By studying large fish, which feed on smaller fish and are living on top of the food chain, scientists can most easily assess antibiotic contamination in the ocean. A recent survey found resistance in all six sites sampled and in all eight fish species studied. (See J. K. Blackburn et al., "Evidence of antibiotic resistance in free-swimming, top-level marine predatory fishes," *Journal of Zoo and Wildlife Medicine* 41 [2010]: 7–15.)

81 **animals fed a drug-free diet:** The idea that antibiotics could be useful as growth promoters was first developed in the 1940s, shortly after their first deployment

to treat infections in both humans and animals: P. R. Moore and colleagues ("Use of sulfasuxidine, streptothricin, and streptomycin in nutritional studies with the chick," *Journal of Biological Chemistry* 165 [1946]: 437–41) are generally credited for this observation. W. J. Visek ("The mode of growth promotion by antibiotics," *Journal of Animal Sciences* 46 [1978]: 1447–69) wrote an outstanding review of the accumulated knowledge about thirty-five years ago; in the light of today's knowledge, the observations seem very accurate. Also see P. Butaye et al., "Antimicrobial growth promoters used in animal feed: effects of less well-known antibiotics on gram-positive bacteria," *Clinical Microbiology Reviews* 16 (2003): 175–88; E. Ozawa, "Studies on growth promotion by antibiotics," *Journal of Antibiotics* 8 (1955): 205–14.

81 **a particularly interesting study from 1963:** M. E. Coates et al., "A comparison of the growth of chicks in the Gustafsson germ-free apparatus and in a conventional environment, with and without dietary supplements of penicillin," *British Journal of Nutrition* 17 (1963): 141–50.

82 **which animals and why:** The Pew Charitable Trust has focused on antibiotic use in food animals. In February 2013 it reported on record-high sales of antibiotics for meat and poultry production. It found that in 2011 nearly 80 percent (30 million) of the nearly 38 million pounds of antibiotics consumed each year in the United States was for animals sold for meat and poultry production. See http://www.pewhealth.org /other-resource/record-high-antibiotic-sales-for-meat-and-poultry-production -85899449119. See also a commentary by former FDA commissioner David Kessler, "Antibiotics and the meat we eat," *New York Times* op-ed page (March 27, 2013).

83 **121 of 132 *Yersinia* samples:** Consumer's Union tested 198 pork chops and ground-pork products purchased at retail in six U.S. cities. Of these, 69 percent were positive for *Yersinia enterocolitica*, an important food-borne pathogen that causes diarrheal and systemic illnesses, most of whose isolates were antibiotic resistant, and of those 39 percent were multiply resistant (*Consumers Reports*, January 2013).

83 **bacteria resistant to antibiotics:** 2011 Retail Meat Report from the National Antimicrobial Resistance Monitoring System. See it at: http://www.fda.gov/downloads /AnimalVeterinary/SafetyHealth/AntimicrobialResistance/NationalAntimicro bialResistanceMonitoringSystem/UCM334834.pdf.

83 **an indication of fecal contamination:** The issues from the 2011 NARMS report are highlighted by the Environmental Working Group in its own report and analysis of the findings: D. Undurraga, "Superbugs invade American supermarkets," http://static .ewg.org/reports/2013/meateaters/ewg_meat_and_antibiotics_report2013.pdf.

83 **forbade the practice in 1999:** M. Casewell et al., "The European ban on growth-promoting antibiotics and emerging consequences for human and animal health," *Journal of Antimicrobial Chemotherapy* 52 (2003): 159–61. The final ban on all growth-promoting antibiotics in the EU went into effect in 2006. But in some countries, farmers circumvented the ban with higher rates of treatments for "infections," which is allowed. Vigilance by regulators is needed.

84 **patterns of antibiotic resistance:** In fall of 2013 a large outbreak of *Salmonella heidelberg* from chickens was one of the latest episodes. The outbreak involved hundreds of people in more than twenty states. Many of the victims were hospitalized because of bloodstream infections caused by these multiply antibiotic-resistant organisms. See CDC, "Multistate outbreak of multidrug-resistant *Salmonella* heidelberg infections linked to Foster Farms brand chicken," http://www.cdc.gov/salmonella/heidel berg-10-13/index.html.

84 **from contact with their animals:** E. M. Harrison et al., "Whole genome sequencing identifies zoonotic transmission of MRSA isolates with the novel *mecA* homologue *mecC*," *EMBO Molecular Medicine* 5 (2013): 509–15.

85 **especially sulfa drugs and tetracycline:** In a November 1990 report to Congress, the General Accounting Office (GAO) indicated that twenty antibiotics were approved for use in dairy cows. It reported results of FDA testing of milk on retail store shelves in several surveys from 1988 to 1990. In all of them, antibiotics, particularly sulfa drugs (including sulfamethazine, which is not approved for use in cattle), were found. Reported rates ranged from 5 to 86 percent, and the GAO questioned whether the FDA tests were sufficiently sensitive. See GAO RCED 91-26, http://www.gao.gov /products/RCED-91-26 and http://www.gao.gov/assets/220/213321.pdf. In China, sulfas and quinolone antibiotics were detected in 40 percent and 100 percent, respectively, of milk sampled in 2011. Levels were reported as low but still widely present (R.-W. Han et al., "Survey of tetracyclines, sulfonamides, sulfamethazine, and quinolones in UHT milk in China market," *Journal of Integrative Agriculture* 12 [2013]: 1300–305).

85 **from treatment plants, and tap water:** C. Xi et al., "Prevalence of antibiotic resistance in drinking water treatment and distribution systems," *Applied and Environmental Microbiology* 75 (2009): 5714–18.

8. MOTHER AND CHILD

88 **the toll of misery mounted relentlessly:** Some doctors prescribed thalidomide to men for its sedative effects. It was safe because men absolutely could not get pregnant. One of these doctors was Jacob Sheskin, my grandmother's first cousin, a dermatologist who cared for patients with leprosy. When he gave thalidomide to several men with advanced leprosy to help them sleep, he observed that one type of their terrible skin lesions improved. He conducted careful clinical trials and proved to a skeptical world that this was true (J. Sheskin, "Thalidomide in the treatment of lepra reactions," *Clinical Pharmacology and Therapeutics* 6 [1965]: 303–6; and J. Sheskin, "The treatment of lepra reaction in lepromatous leprosy. Fifteen years' experience with thalidomide," *International Journal of Dermatology* 6 [1980]: 318–22). Sheskin was a clinician and did not understand the basis for thalidomide's action, but later others did, and they went on to extend its uses. Today thalidomide and a family of related drugs are used in cancer therapy as a mainstay for certain conditions,

including multiple myeloma and other tumors. If someone predicted this fifty years ago, everyone would have assumed it to be a very sick form of joke.

89 **Did not improve pregnancy outcomes in the least:** From the early 1940s through the 1960s, diethylstilbestrol (DES) was prescribed for pregnant women to reduce the risk of pregnancy complications and losses. However, beginning in the early 1950s, studies began to appear in the obstetrics literature indicating that DES was not effective in promoting better pregnancy outcomes. For example, a widely cited clinical trial that was performed in Chicago showed no improvement in adverse pregnancy outcomes in women who were randomly assigned to receive DES or to serve as controls. (W. J. Dieckmann et al., "Does the administration of diethylstilbestrol during pregnancy have therapeutic value?" *American Journal of Obstetrics and Gynecology* 66 [1953]: 1062–81.) By the time that DES usage stopped in the late 1960s, millions of pregnant women (and thus their babies) had received the drug. See also R. J. Apfel and S. M. Fisher, *To Do No Harm: DES and the Dilemmas of Modern Medicine* (New Haven: Yale University Press, 1986).

89 **clear-cell adenocarcinoma of the vagina:** A. L. Herbst et al., "Adenocarcinoma of the vagina: association of maternal stilbestrol therapy with tumor appearance in young women," *New England Journal of Medicine* 284 (1971): 878–81.

89 **(33.3 percent vs. 15.5 percent):** R. Hoover et al., "Adverse health outcomes in women exposed in utero to diethylstilbestrol," *New England Journal of Medicine* 365 (2011): 1304–14. As indicated on its website, "the DES Follow-Up Study investigates the long-term health consequences associated with exposure to diethylstilbestrol (DES). Since 1992, the National Cancer Institute in collaboration with research centers throughout the United States has been conducting the DES Follow-Up Study of more than 21,000 mothers, daughters, and sons."

91 **is coming under question:** As discussed in the chapter 3 notes, recently investigators have pointed out that in many animal species, the transfer of microbes from mother to child begins before birth, while their baby is still in the womb (Funkhauser and Bordenstein, "Mom knows best"). There isn't much information yet about humans, but studies should address this in the next few years. If this occurs, then the importance of antibiotic use during pregnancy may rise.

91 **women who have been studied:** O. Koren et al., "Host remodeling of the gut microbiome and the metabolic changes during pregnancy," *Cell* 150 (2012): 470–80. This is the first part of the study in Ruth Ley's lab that is discussed below.

93 **rapidly colonizes the mother's skin:** M. G. Domínguez-Bello et al., "Delivery mode shapes the acquisition and structure of the initial microbiota across multiple body habitats in newborns," *Proceedings of the National Academy of Sciences* 107 (2010): 11971–75.

96 **foundation of microbes:** In a study of the gut microbiota of healthy people in three places—the United States, Malawi, and Venezuela (Amerindians)—Yatsunenko and her colleagues, including my wife, Gloria, catalogued which microbes were present across people of all ages (see chapter 1). Initially I was surprised, but the more I thought

about it, the more sense it made. This was consistent with my hypotheses about the importance of the early-life microbiota.

97 **the rate is about 4 percent:** In Järna, a Swedish community near Stockholm, families try to sustain as natural a lifestyle as possible. They minimize antibiotic use, and virtually all of their babies are breast fed. They try to have C-sections only when absolutely required: their rate, at 4 percent, is lower than the rest of Sweden (about 17 percent), and a lot lower than that of the United States (32 percent). See J. S. Alm et al., "An anthroposophic lifestyle and intestinal microflora in infancy," *Pediatric Allergy and Immunology* 13 (2002): 402–11.

97 **to one in three births in 2011:** In 1981, of one hundred women coming to the hospital to give birth, the rates in nineteen industrialized countries ranged from 5 percent in Czechoslovakia to 18 percent in the United States. (See F. C. Notzon et al., "Comparisons of national Cesarean-section rates," *New England Journal of Medicine* 316 [1987]: 386–89.) More recently, the rates in the United States climbed to 30.5 percent between 2002 and 2008. (See J. Zhang et al., "Contemporary Cesarean delivery practice in the United States," *American Journal of Obstetrics and Gynecology* 203 [2010]: 326.e1–10.) For births in 2011, the CDC reported a national rate of 32.8 percent, a more than 80 percent increase in thirty years from our already high rate.

97 **13 percent in the Netherlands:** According to the World Health Organization (2008), the highest rates in the world are in countries like Brazil (46 percent), Iran (42 percent), and the Dominican Republic (42 percent). The lowest country is the Netherlands (about 13 percent), and Scandinavia, in general, has rates much lower than that of the rest of the world. Is the medical care in Brazil, Iran, and the Dominican Republic more advanced than it is in northern Europe, or are other factors at play?

101 **and all will get intravenous penicillin:** Unless the mother is allergic to penicillin, in which case another antibiotic will be substituted.

101 **acquired from his or her mother:** "Prevention of Perinatal Group B Streptococcal Disease," Revised Guidelines from CDC, 2010, MMWR, *Recommendations and Reports* 59(RR10): (Nov. 19, 2010): 1–32.

For our review of the subject of current antibiotic use in pregnancy, see W. J. Ledger and M. J. Blaser, "Are we using too many antibiotics during pregnancy?" *British Journal of Obstetrics and Gynecology* 120 (2013): 1450–52; I. A. Stafford et al., "Efficacy of maternal and neonatal chemoprophylaxis for early-onset group B streptococcal disease," *Obstetrics and Gynecology* 120 (2012): 123–29. Although overall national rates of early-onset sepsis have diminished substantially, rates of early-onset GBS sepsis were unchanged after thirteen years of prophylaxis at one major medical center, reflecting a host of accumulated problems.

102 **get the procedure:** Episiotomy rates vary widely from country to country (see I. D. Graham et al., "Episiotomy rates around the world: an update," *Birth* 32 [2005]: 219–23). For an older, very comprehensive review, see G. Carroli and J. Belizan,

"Episiotomy for vaginal birth," *Cochrane Database of Systematic Reviews* 3, no. CD000081 (2007): DOI: 10.1002/14651858.CD000081. Also see F. Althabe et al., "Episiotomy rates in primiparous women in Latin America: hospital-based descriptive study," *British Medical Journal* 324 (2002): 945–46.

102 **first silver nitrate:** Dr. Albert Barnes developed a dilute solution of silver nitrate around the turn of the twentieth century. It was called Argyrol and was used to treat gonorrheal eye infections that lead to blindness. He sold his company in 1929 for millions on the eve of the stock market crash. The proceeds from Argyrol formed the foundation for the renowned Barnes Foundation art collection in Philadelphia.

103 **among the millions of births a year:** We now screen pregnant women for HIV so that with proper prevention we can almost eliminate the risk of its transmission to the baby.

9. A FORGOTTEN WORLD

105 **depending on context:** Theodor Rosebury, a student and researcher of the oral microbiota beginning in the 1930s, had great insight into the biological relationships that we have with our residential organisms. His seminal works include: *Microorganisms Indigenous to Man* (New York: McGraw Hill, 1962) and *Life on Man* (London: Seeker and Warburg, 1969). In 1962 he coined the neologism *amphibiosis.* Today's scientists like the concept so much that they use it with a more modern name. They call the microbes pathobionts rather than the amphibionts originally described by Rosebury. But since it is the same idea, I give Rosebury credit and use his terms throughout.

106 **in the continent's jungles and highlands:** From Venezuelan patients who underwent upper gastrointestinal endoscopy, María Gloria Domínguez Bello (later my wife) and her colleagues in Venezuela obtained gastric biopsies from those living in urban areas near the coast and from deep in the interior in Puerto Ayacucho, the capital of Amazonas State. The *H. pylori* strains (that were isolated in pure culture by Chandra Ghose, a graduate student in my lab) from the biopsies of some of the Amerindian patients in Puerto Ayacucho had genetic signatures that were closely related to strains from present-day people in China and Japan. In contrast, the strains isolated from the coastal patients had the signatures of present-day Europeans and Africans. The most parsimonious explanation for these findings is that the ancestors of the Amerindians had East Asian strains of *H. pylori* in their stomachs when they crossed the Bering Strait and that the descendants of those people and their *H. pylori* strains flourished until Columbus and the European conquest. With the decimation of the Amerindians and the introduction of *H. pylori* strains arriving in the stomachs of Europeans and their African slaves, there were few remaining Amerindian strains in coastal areas. But deep in the interior, these strains persisted, and the secret of their ancestry could be revealed by DNA sequencing. (See C. Ghose et al., "East Asian genotypes of *Helicobacter pylori* strains in Amerindians provide evidence for its ancient human carriage," *Proceedings of the National Academy of Sciences* 99 [2002]: 15107–11.) Subsequently, we worked with an international team that made

sense of a worldwide collection of *H. pylori* isolates to understand how the organisms spread around the world in the past 58,000 years. (See D. Falush et al., "Traces of human migration in *Helicobacter pylori* populations," *Science* 299 [2003]: 1582–85.) In later studies, Gloria and her colleagues deciphered the whole genome sequence of one of the strains from Puerto Ayacucho, showing its uniqueness in the *H. pylori* universe. (See S. P. Mane et al., "Host-interactive genes in Amerindian *Helicobacter pylori* diverge from their old world homologs and mediate inflammatory responses," *Journal of Bacteriology* 192 [2010]: 3078–92.)

110 **organisms from fecal specimens:** The methods used for isolation of *Campylobacter* were developed by Martin Skirrow, a clinical microbiologist in Worcester, England. It was his paper (M. Skirrow, "Campylobacter enteritis: a new disease," *British Medical Journal* 2 [1977]: 9–11), which I read in July 1977 shortly after I had been caring for the patient with *C. fetus* infection (see chapter 1), that brought me into the field of medical research. Later we modified Skirrow's medium (M. J. Blaser et al., "Campylobacter enteritis: clinical and epidemiologic features," *Annals of Internal Medicine* 91 [1979]: 179–85) to improve isolation of these fastidious organisms.

110 **never bothered again by *H. pylori*:** B. J. Marshall et al., "Attempt to fulfil Koch's postulates for pyloric campylobacter," *Medical Journal of Australia* 142 (1985): 436–39.

111 **the same relationships in their own studies:** B. J. Marshall et al., "Prospective double-blind trial of duodenal ulcer relapse after eradication of *Campylobacter pylori*," *Lancet* 2 (1988): 1437–42. An Irish group (J. G. Coghlan et al., "Campylobacter pylori and recurrence of duodenal ulcers—a 12-month follow-up study," *Lancet* 2 [1987]: 1109–11) published similar findings more than a year earlier but, despite more than five hundred citations, their paper was mostly forgotten, and Marshall et al. received most of the scientific credit. Later studies in the United States (D. Y. Graham et al., "Effect of treatment of *Helicobacter pylori* infection on the long-term recurrence of gastric or duodenal ulcer: a randomized, controlled study," *Annals of Internal Medicine* 116 [1992]: 705–8) and in Austria (E. Hentschel et al., "Effect of ranitidine and amoxicillin plus metronidazole on the eradication of *Helicobacter pylori* and the recurrence of duodenal ulcer," *New England Journal of Medicine* 328 [1993]: 308–12) cemented the finding that antibiotic treatments that eradicated *H. pylori* markedly improved the natural history of peptic ulcer disease, often leading to a cure.

112 **based on their having antibodies to the organism:** Guillermo received his doctoral degree based on studies we did together exploring the nature of the antigens of *C. jejuni* and *C. fetus* strains. By 1985 I was convinced that this new "campylobacter" could be medically important, so we began to apply the same biochemical and immunological approaches to it. (See G. I. Pérez-Pérez and M. J. Blaser, "Conservation and diversity of *Campylobacter pyloridis* major antigens," *Infection and Immunity* 55 [1987]: 1256–63; G. I. Pérez-Pérez et al., "*Campylobacter pylori* antibodies in humans," *Annals of Internal Medicine* 109 [1988]: 11–17.)

113 **a very good lead:** We were particularly interested in comparing people with ulcers
with those who had gastritis only. Among the seventy-four patients who had gastritis
(meaning inflammation of the stomach), about 60 percent had antibodies to the CagA
protein. But each of the thirty-one patients who had duodenal ulcers had these antibodies. (See T. L. Cover et al., "Characterization of and human serologic response to
proteins in *Helicobacter pylori* broth culture supernatants with vacuolizing cytotoxin
activity," *Infection and Immunity* 58 [1990]: 603–10.) For the first time, we had a blood
test that could highlight people who were at high risk for getting ulcers. About a year
and a half later, a group in England led by Jean Crabtree ("Mucosal IgA recognition
of *Helicobacter pylori* 120 kDa protein, peptic ulceration, and gastric pathology," *Lancet*
338 [1991]: 332–35) identified the same protein with the same proportions of people having antibodies when they had gastritis (~60 percent) and ulcers (100 percent)
as we did. At that point, I was certain we had identified a critical *H. pylori* protein—
two independent groups, across the ocean, with nearly identical observations—not
a chance event. The confirmation makes a discovery a real discovery.

113 **for cytotoxin-associated gene:** As often happens in science, a second group, from
the Biocene Company in Siena, Italy, was following a similar strategy. Even though
we recognized the association with ulcers earlier and cloned the gene earlier, once
they got started, they more rapidly uncovered many of the same associations. By
chance we learned that they had identified the same gene and that they had given it
a different name. Guided by a shared spirit of scientific collaboration, we eventually
agreed on a common name, CagA, because those strains produced unusually high
levels of the cytotoxin that injures human cells (cytotoxin-associated gene A). This
collaboration saved the field from the uncertainty and divisiveness that occur when
there are two names for the same thing. (See M. Tummuru et al., "Cloning and
expression of a high-molecular-mass major antigen of *Helicobacter pylori*: evidence of
linkage to cytotoxin production," *Infection and Immunity* 61 [1993]: 1799–809; A.
Covacci et al., "Molecular characterization of the 128-kDa immunodominant antigen of *Helicobacter pylori* associated with cytotoxicity and duodenal ulcer," *Proceedings of
the National Academy of Sciences* 90 [1993]: 5791–95.)

114 **we discovered and named VacA:** T. L. Cover et al., "Divergence of genetic sequences
for the vacuolating cytotoxin among *Helicobacter pylori* strains," *Journal of Biological Chemistry* 269 (1994): 10566–73; and T. L. Cover and M. J. Blaser, "Purification and characterization of the vacuolating toxin from *Helicobacter pylori*," *Journal of Biological Chemistry*
267 (1992): 10570–75, described the discovery of the protein, which we called VacA,
and then we called the gene *vacA*. VacA was discovered as a toxin, but I now think of
it as a signaling molecule, a way in which *H. pylori* tells the host what it wants the host
to do. One effect of VacA is to tone down the immune response of T-cells, as a way of
ensuring its own survival. (See B. Gebert et al., "*Helicobacter pylori* vacuolating cytotoxin
inhibits T lymphocyte activation," *Science* 301 [2003]: 1099–1102.) If it is too toned
down, there might not be enough inflammation for *H. pylori*, and thus not enough

nutrients. So it has to strike a balance. Many years ago, Tim and I had the notion that CagA is the accelerator and VacA is the brake. It stills looks like a good idea.

114 **But two years later:** In 1989 we published an article in the *New England Journal of Medicine* on the relationship of *H. pylori* to gastritis (C. P. Dooley et al., "Prevalence of *Helicobacter pylori* infection and histologic gastritis in asymptomatic persons," *New England Journal of Medicine* 321 [1989]: 1562–66), again confirming the utility of our blood test. After he read that article, Dr. Nomura wrote to me. We communicated mostly by mail and sometimes by telephone. Although we worked closely on a number of studies that were very important to both of us, and worked very well together, we did not actually meet in person for about ten years!

115 **strains had double the risk:** Four papers all published in 1991 showed strong associations of having *H. pylori* and developing gastric cancer: J. Parsonnet et al., "*Helicobacter pylori* infection and the risk of gastric carcinoma," *New England Journal of Medicine* 325 (1991): 1127–31; A. Nomura et al., "*Helicobacter pylori* infection and gastric carcinoma among Japanese Americans in Hawaii," *New England Journal of Medicine* 325 (1991): 1132–36; D. Forman et al., "Association between infection with *Helicobacter pylori* and risk of gastric cancer: evidence from a prospective investigation," *British Medical Journal* 302 (1991): 1302–5; and N. J. Talley et al., "Gastric adenocarcinoma and *Helicobacter pylori* infection," *Journal of the National Cancer Institute* 83 (1991): 1734–39. Later we showed that having a *cagA*+ strain just about doubled the risk of gastric cancer development (M. J. Blaser et al., "Infection with *Helicobacter pylori* strains possessing *cagA* is associated with an increased risk of developing adenocarcinoma of the stomach," *Cancer Research* 55 [1995]: 2111–15) and of its precursor, chronic atrophic gastritis (E. J. Kuipers et al., "*Helicobacter pylori* and atrophic gastritis: importance of the cagA status," *Journal of the National Cancer Institute* 87 [1995]: 1777–80).

115 **"the only good *Helicobacter* pylori is a dead one":** D. Y. Graham, "The only good *Helicobacter pylori* is a dead *Helicobacter pylori*," *Lancet* 350 (1997): 70–71.

116 **This ancient organism:** Evidence that *H. pylori* is ancient: D. Falush et al., "Traces of human migration in *Helicobacter pylori* populations," *Science* 299 (2003): 1582–85; B. Linz et al., "An African origin for the intimate association between humans and *Helicobacter pylori*," *Nature* 445 (2007): 915–18; Y. Moodley et al., "The peopling of the Pacific from a bacterial perspective," *Science* 323 (2009): 527–30; S. Breurec et al., "Evolutionary history of *Helicobacter pylori* sequences reflect past human migrations in Southeast Asia," *PLOS ONE* 6 (2011): e22058: 1–10; and Y. Moodley et al., "Age of the association between *Helicobacter pylori* and man," *PLOS Pathogens* 8 (2012): e1002693: 1–16.

118 **whether his mother has it:** J. Raymond et al., "Genetic and transmission analysis of *Helicobacter pylori* strains within a family," *Emerging Infectious Diseases* 10 (2004): 1816–21.

120 **the human stomach has changed markedly:** M. J. Blaser, "*Helicobacter pylori* eradication and its implications for the future," *Alimentary Pharmacology and Therapeutics* 11, suppl. 1 (1997): 103–7; "Not all *Helicobacter pylori* strains are created equal: should

all be eliminated?" 349 *Lancet* (1997): 1020–22; "Helicobacters are indigenous to the human stomach: duodenal ulceration is due to changes in gastric microecology in the modern era," *Gut* 43 (1998): 721–27; "In a world of black and white, *Helicobacter pylori* is gray," *Annals of Internal Medicine* 130 (1999): 695–97.

121 **they're safely across:** M. J. Blaser and D. Kirschner, "The equilibria that allow bacterial persistence in human hosts," *Nature* 449 (2007): 843–49.

10. HEARTBURN

123 **have symptoms every day:** G. M. Eisen et al., "The relationship between gastroesophageal reflux and its complications with Barrett's esophagus," *American Journal of Gastroenterology* 92 (1997): 27–31; and H. B. El-Serag, "Time trends of gastroesophageal reflux disease: a systematic review," *Clinical Gastroenterology and Hepatology* 5 (2007): 17–26.

126 **progress to the form of cancer called adenocarcinoma:** J. Lagergren et al., "Symptomatic gastroesophageal reflux as a risk factor for esophageal adenocarcinoma," *New England Journal of Medicine* 340 (1999): 825–31.

126 **first identified in 1950:** In 1950, Norman Barrett, an English surgeon, reported finding abnormal tissue in the esophagus. We now call that abnormality Barrett's esophagus, and he became Sir Norman Barrett.

126 **in the past three decades:** The incidence of adenocarcinoma of the esophagus is rising, and not just because of better surveillance and reporting. (See H. Pohl and H. G. Welsh, "The role of overdiagnosis and reclassification in the marked increase of esophageal adenocarcinoma incidence," *Journal of the National Cancer Institute* 97 [2005]: 142–46.)

127 **eight times more likely:** J. J. Vicari et al., "The seroprevalence of *cagA*-positive *Helicobacter pylori* strains in the spectrum of gastroesophageal reflux disease," *Gastroenterology* 115 (1998): 50–57; and M. F. Vaezi et al., "*CagA*-positive strains of *Helicobacter pylori* may protect against Barrett's esophagus," *American Journal of Gastroenterology* 95 (2000): 2206–11. Other more recent GERD/Barrett's studies include: D. Corley et al., "*Helicobacter pylori* infection and the risk of Barrett's oesophagus: a community-based study," *Gut* 57 (2008): 727–33; and L. A. Anderson et al., "Relationship between *Helicobacter pylori* infection and gastric atrophy and the stages of the oesophageal inflammation, metaplasia, adenocarcinoma sequence: results from the FIN-BAR case-control study," *Gut* 57 (2008): 734–39. All show inverse associations, with the strongest data from the Corley study showing that people with *cagA*+ strains have a 92 percent reduction in the risk of developing Barrett's esophagus.

128 **double the rate of esophageal disease:** J. Labenz et al., "Curing *Helicobacter pylori* infection in patients with duodenal ulcer may provoke reflux esophagitis," *Gastroenterology* 112 (1997): 1442–47.

128 **colleagues from around the world:** W. H. Chow et al., "An inverse relation between *cagA*+ strains of *Helicobacter pylori* infection and risk of esophageal and gastric cardia adenocarcinoma," *Cancer Research* 58 (1998): 588–90; R. Peek et al., "The role of

Helicobacter pylori cagA+ strains and specific host immune responses on the development of premalignant and malignant lesions of the gastric cardia," *International Journal of Cancer* 82 (1999): 520–24; R. J. L. F. Loffeld et al., "Colonization with *cagA*-positive *H. pylori* strains inversely associated with reflux oesophagitis and Barrett's oesophagitis," *Digestion* 62 (2000): 95–99; and F. Kamangar et al., "Opposing risks of gastric cardia and noncardia gastric adenocarcinomas associated with *Helicobacter pylori* seropositivity," *Journal of the National Cancer Institute* 98 (2006): 1445–52.

129 **and are more damaging:** Two studies that were presented at scientific congresses in the late 1990s were very instructive. In the Eurogast Study (P. M. Webb et al., "Gastric cancer, cytotoxin-associated gene A-positive *Helicobacter pylori*, and serum pepsinogens: an international study," *Gastroenterology* 116 [1999]: 269–76), 2,850 patients in thirteen countries who underwent upper gastrointestinal endoscopy were examined with stomach biopsies, and assessed for blood levels of proteins that are made in the stomach. As was by then expected, people with *H. pylori* had a higher rate of showing a blood-protein (pepsinogen) ratio indicative of atrophic changes than those without the organism, and those with *cagA+* strains had even more altered blood-protein ratios than those with *cagA*-negative strains. In a contemporaneous study, Y. Yamaji et al. ("Inverse background of *Helicobacter pylori* antibody and pepsinogen in reflux oesophagitis compared with gastric cancer: analysis of 5732 Japanese subjects," *Gut* 49 [2001]: 335–40) showed that in Japan people who had reflux had a pattern of gastric tissue changes and protein production that was the inverse of those with stomach cancer. As the signs of atrophic gastritis—a precursor to gastric cancer—rose, the prevalence of reflux fell. These two studies, done with thousands of patients, provide support for the mixed roles of *H. pylori* vis-à-vis diseases of the stomach and the esophagus.

11. TROUBLE BREATHING

133 **we understood how these strains operate:** Mechanism of action in CagA+ strains. In 1995 we published the first evidence that *H. pylori* had a Type-IV secretion system that could transport *H. pylori* products into the cells lining the stomach wall, but we did not know what was being exported (M. Tummuru et al., *Helicobacter pylori picB*, a homologue of the *Bordetella pertussis* toxin secretion protein, is required for induction of IL-8 in gastric epithelial cells," *Molecular Microbiology* 18 [1995]: 867–76). By 2000, several groups (most notably S. Odenbreit et al., "Translocation of *Helicobacter pylori* CagA into gastric epithelial cells by Type IV secretion," *Science* 287 [2000]: 1497–1500; as well as A. Covacci and R. Rappuoli, "Tyrosine-phosphorylated bacterial proteins: Trojan horses for the host cell," *Journal of Experimental Medicine* 191 [2000]: 587–92) proved that there was a Type-IV system, and that the material it injected was none other than the CagA protein that we (and Covacci) had found a decade earlier, in our case, probing the library of *H. pylori* genes with my serum (see chapter 9). No wonder that I made antibodies to CagA; the *H. pylori* strain that I was carrying was injecting the protein into my stomach wall every day for years.

134 **at the age of seven:** A. L. Kozyrskyj et al., "Increased risk of childhood asthma from antibiotic use in early life," *Chest* 131 (2007): 1753–59.

134 **in May of that year:** M. E. Fernández-Beros, L. Rogers, G. I. Pérez-Pérez, W. Hoerning, M. J. Blaser, and J. Reibman, "Seroprevalence of *Helicobacter pylori* is associated with later age of onset of asthma in urban adults," abstract presented in May 2005 at the American Thoracic Society Annual Meeting in San Diego, CA.

135 **the subject's *H. pylori* status:** In the late 1990s Guillermo had run tests on more than eleven thousand people in that study as part of a government contract. He had just received little numbered tubes of serum without any knowledge of any of the characteristics of the people they came from. He was completely blinded, and some of the samples were in fact deliberate duplicates to see how reproducible the assays were. His results were beautifully reproducible, which made the sponsors (and us) very pleased. After the paper was published (J. E. Everhart et al., "Seroprevalence and ethnic differences in *Helicobacter pylori* infection among adults in the United States," *Journal of Infectious Diseases* 181 [2000]: 1359–63), eventually a lot of the data from NHANES III became publicly available in complicated tables and spreadsheets that qualified statisticians could examine. We had earlier examined the relationship of *H. pylori* and obesity using NHANES III (I. Cho et al., "*Helicobacter pylori* and overweight status in the United States: data from the Third National Health and Nutrition Examination Survey," *American Journal of Epidemiology* 162 [2005]: 579–84) and did not find any relationship, but we had practice using the complex NHANES III data set.

135 **my hypothesis was correct:** J. Reibman et al., "Asthma is inversely associated with *Helicobacter pylori* status in an urban population," *PLOS ONE* 3 (2008): e4060: 1–6; and Y. Chen and M. J. Blaser, "Inverse associations of *Helicobacter pylori* with asthma and allergies," *Archives of Internal Medicine* 167 (2007): 821–27.

137 **studies show consistent results:** Y. Chen and M. J. Blaser, "*Helicobacter pylori* colonization is inversely associated with childhood asthma," *Journal of Infectious Diseases* 198 (2008): 553–60.

138 **the modern *H. pylori*–free stomach:** R. Rad et al., "CD25+/Foxp3+ T cells regulate gastric inflammation and *Helicobacter pylori* colonization in vivo," *Gastroenterology* 131 (2006): 525–37; and K. Robinson et al., "*Helicobacter pylori*–induced peptic ulcer disease is associated with inadequate regulatory T cell responses," *Gut* 57 (2008): 1375–85.

139 **as Mueller and her colleagues showed in mice:** I. C. Arnold et al., "*Helicobacter pylori* infection prevents allergic asthma in mouse models through the induction of regulatory T cells," *Journal of Clinical Investigation* 121 (2011): 3088–93; and M. Oertli et al., "DC-derived IL-18 drives Treg differentiation, murine *Helicobacter pylori*–specific immune tolerance, and asthma protection," *Journal of Clinical Investigation* 122 (2012): 1082–96.

141 **in the next twenty-one years:** A. Nomura et al., "*Helicobacter pylori* infection and the risk for duodenal and gastric ulceration," *Annals of Internal Medicine* 120 (1994): 977–81.

141 **in the other organisms present and their distribution:** M. J. Blaser, "Helicobacters are indigenous to the human stomach: duodenal ulceration is due to changes in gastric microecology in the modern era," *Gut* 43 (1998): 721–27.

12. TALLER

145 **one of my colleagues at NYU:** Lewis Goldfrank, MD, chair of the Department of Emergency Medicine at NYU Langone Medical Center and at Bellevue Hospital Center.

146 **a rural community in Guatemala:** L. Mata, *The Children of Santa Maria Cauque: a Prospective Field Study of Health and Growth* (Cambridge, MA: MIT Press, 1978).

146 **the development of adult height:** A. S. Beard and M. J. Blaser, "The ecology of height: the effect of microbial transmission on human height," *Perspectives in Biology and Medicine* 45 (2002): 475–98.

146 **acquired in the first few years of life:** Even today, we don't fully know how *H. pylori* is acquired in early life. We know that having an *H. pylori*–negative mother strongly reduces the risk of a child acquiring the organism, but thus far it has not been found in the vagina, and even in communities in which nearly all of the mothers are positive, we rarely detect it in their children before the age of one year. Either it was there all along and was suppressed or it is actually acquired later, still from the mother or from siblings, father, or friends (in day care and school). Or it is possible that for one hundred positive children, it is a mixture of all of these routes, but it remains a mystery. We know that it is not from the family dog, because dogs don't carry *H. pylori*; they have their own helicobacters.

147 **the same fecal-oral route:** This is a mechanism for transmitting microbes in the feces of one person into the mouth of another. Food, water, hands may be intermediaries. Infectious diseases such as polio, hepatitis A, and typhoid fever are transmitted this way.

147 **the hormones ghrelin and leptin:** C. U. Nwokolo et al., "Plasma ghrelin following cure of *Helicobacter pylori*," *Gut* 52 (2003): 637–40; and F. François et al., "The effect of *H. pylori* eradication on meal-associated changes in plasma ghrelin and leptin," *BMC Gastroenterology* 11 (2011): 37.

147 **keeping records of it:** Beard and Blaser, "The ecology of height."

151 **bad things can happen:** M. J. Blaser and D. Kirschner. "The equilibria that allow bacterial persistence in human hosts," *Nature* 449 (2007): 843–49.

13. . . . AND FATTER

152 **visiting scholars from around the world:** But where did the money come from to do the work? There is a paradox in the way that medical science is funded in the United States and elsewhere. In order to get a grant, you must have "preliminary data" that support the idea to be tested, to see whether it may be feasible. But how do you do the feasibility studies without the money that the grant provides? It is a

catch-22. I was fortunate at the time to have the means to fund a new idea. First, because I had a number of ongoing research projects, I had amassed over the years equipment and supplies that we could use for a new project. Then, being at a university, there were students and trainees who were looking for a new project, a way to develop their own career track. Ilseung came my way for that reason. I also had received support from several philanthropists that was not exactly earmarked; I could spend it at my discretion. We often say that discretionary money is worth double because of its flexibility. With that support, I could make a commitment to Laurie when she was looking for a lab in which to do graduate work. Finally, there is luck, too. A colleague told me about a neighbor who was looking to work in the lab for a summer. He also told me that she was a student at Princeton. I felt that was a good predictor that she would have something on the ball, and when I met Yael that premise was immediately confirmed. In the United States, the conduct of science is very entrepreneurial, and after good ideas hard work is the sine qua non of success.

153 **"What's happening to their body composition?":** Ilseung later applied for and received an NIH-supported grant from NYU's Clinical and Translational Science Institute (CTSI) that allowed him to conduct this research. CTSI director Dr. Bruce Cronstein, who was a member of a mentoring committee to help Ilseung solve research problems, has studied the metabolism of bone for many years and had a DEXA machine that he used for his own mouse studies. He made the suggestion that opened up new vistas for us. In science, it also "takes a village" to get things done.

156 **the adipose tissue, where fat is stored:** I. Cho et al., "Antibiotics in early life alter the murine colonic microbiome and adiposity," *Nature* 488 (2012): 621–26. This was Ilseung's major work in the lab and involved him and twelve other scientists—biochemists, animal experimenters, informaticists, gene expression analysts—each contributing to a different aspect of the work. But Ilseung's patient studies and the sixteen months we spent conducting new experiments and clarifying our work for the anonymous reviewers and the editors of *Nature* paid off, and the paper was finally published, more than five years after the work had started.

156 **were altered from the get-go.:** L. Cox et al., "Altering the intestinal microbiota during a critical developmental window has lasting metabolic consequences," *Cell* 158 (2014); 705–21. This recently published paper provides the details for Laurie's experiments cited on pages 156–162.

157 **But when they are conventionalized:** The germ-free state is artificial; there are no germ-free animals except within specialized laboratories. When germ-free animals have their microbiota restored, it is said that they have been "conventionalized" back to the usual (natural) state.

157 **from the normal-weight mouse donor:** In Jeff's experiments, the germ-free mice can be used as living test tubes that can react to the newly introduced microbiota. ("Germ-free" means that the playing field is even and clean, tabula rasa.) (P. Turnbaugh et al., "A core gut microbiome in obese and lean twins," *Nature* 457 [2009]: 480–84.)

161 **can make a big difference:** Laurie later did studies to assess how faithfully we had transferred the microbes from their original hosts to the recipients. The DNA sequencing results showed that we did amazingly well. Even microbes for which we had the evidence of only a single sequence in our snapshot of what was there were well represented in the recipients. Thus we had confidence that the germ-free mice were colonized with what actually was in the STAT or control mice. Interestingly, the community of STAT microbes didn't do as well in their new hosts as the community of untreated microbes. The population was less resilient, and were less resistant to invasion by new species. In this second generation of STAT mice, the microbiota were punier, and I worry about this. See chapter 15, "Antibiotic Winter."

163 **all the antibiotics prescribed for American children:** We decided to study the two classes of antibiotics most commonly used for human children. The first, beta-lactams, include penicillin, amoxicillin, Augmentin (amoxicillin with a second compound to inhibit bacterial enzymes that would inactivate it), and cephalosporins. Amoxicillin is the number-one drug prescribed to young children in the United States and in most developed countries. In 2010 there were nearly 23 million courses of amoxicillin or Augmentin prescribed to children in the United States and more than 6.5 million of those courses were for children under the age of two. (G. Chai et al., "Trends of outpatient prescription drug utilization in U.S. children, 2002–2010," *Pediatrics* 130 [2012]: 23–31.) That averages to nearly one course of those amoxicillin-based antibiotics per young child per year. Macrolides are the second class of antibiotics used for young children. The best known is erythromycin, available for more than fifty years, but in the past twenty years, longer-acting and broader-spectrum agents have been used, including clarithromycin and azithromycin (the Z-pak, which has benefited from a marketing strategy as good as any ever employed). In 2010 U.S. children received more than 10 million courses of azithromycin, which has become the most widely prescribed antibiotic in the United States. It was so expensive that anyone who bought it was likely to use it. (It is now off-patent and the price has dropped.) The tylosin that we used is the macrolide that can most easily and inexpensively be used in mice and for which there is an extensive literature that helped us figure out the correct doses.

163 **the drug that most promotes the recent increases in human height:** The increase in human height began before antibiotics were discovered, at least in the Western countries. But these experiments (both STAT and PAT) indicate that antibiotics—and it is not limited to a single type—affect microbiome composition (see below) and can affect early-life bone development. Certainly this could be part of the story and could explain why the recent height increases in China are recapitulating in forty years the increases that took one hundred years in Europe and the United States.

164 **colleagues at Washington University in St. Louis:** We worked closely together with Drs. Erica Sodergren and George Weinstock, who run a major genome-sequencing center at Washington University in St. Louis. Once we received the sequence infor-

mation from them, Alex Alekseyenko, an NYU faculty member who is an expert in bioinformatics, decoded and deconstructed the data, and then analyzed it.

164 **passed on to them by their mother:** We could not find evidence for the presence of many of the microbes seen in the mothers and in the control mice. Either they had been permanently eliminated or they were still present but in low numbers, below our ability to detect them—in which case, the bacteria that bloomed under the influence of the tylosin regimen were still suppressing them—long after the tylosin was gone. This can happen because they get such an advantage in early life—a "founder effect"—that they are able to sustain their increased numbers.

165 **big studies now under way:** Through Dr. Ernst Kuipers, my former postdoctoral trainee and now longtime friend, we were working with a group in the Netherlands to address this. A large cohort, including more than ten thousand mothers and their newborn children in Rotterdam, have been enrolled in the kind of study that could provide answers to important questions in development, but it will take several years for the kids to get old enough to have any reasonable outcome data. The United States is in the early stages of the National Children's study, whose goal is to enroll up to one hundred thousand children, get lots of information, and see what kind of outcomes—especially asthma, obesity, and diabetes—they develop. The results of that study also will become available years from now.

165 **Avon Longitudinal Study of Parents and Children (ALSPAC) study in Britain:** Drs. Leo Trasande and Jan Blustein, NYU faculty members who work primarily in pediatrics and health policy, respectively, are expert epidemiologists. They both found the ALSPAC Study (J. Golding et al., "ALSPAC—the Avon Longitudinal Study of Parents and Children, I. Study methodology," *Paediatric and Perinatal Epidemiology* 15 [2001]: 74–87) and led the analyses (L. Trasande et al., "Infant antibiotic exposures and early-life body mass," *International Journal of Obesity* 37 [2013]: 16–23; J. Blustein et al., "Association of caesarian delivery with child adiposity from age 6 weeks to 15 years," *International Journal of Obesity* 37 [2013]: 900–906).

166 **may also contribute to the risk:** A Boston study of 1,255 mother-child pairs (S. Y. Huh et al., "Delivery by caesarean section and risk of obesity in preschool children: a prospective cohort study," *Archives of the Diseases of Childhood* 97 [2012]: 610–16) found a significantly increased obesity risk in offspring born by C-section. A Canadian study (K. Flemming et al., "The association between caesarean section and childhood obesity revisited: a cohort study," *Archives of the Diseases of Childhood* 98 [2013]: 526–32) showed that C-section was a risk factor overall, but when they controlled for excessive maternal weight it dropped out. Similarly in our ALSPAC study (J. Blustein et al.), nearly all of the risk was in babies born of mothers who already were overweight. There are many potential explanations for this; one is that the overweight mothers already have a depleted microbiota and C-section adds to the problem in the next generation. In Brazil, where in 2009 the C-section rate went above 50 percent, meaning that more than half the 3 million births in that country were

by C-section, two studies showed different results. H. A. S. Goldani et al. ("Cesarean delivery is associated with an increased risk of obesity in adulthood in a Brazilian birth cohort study," *American Journal of Clinical Nutrition* 93 [2011]: 1344–47), studying a 1978 birth cohort twenty-three to twenty-five years later, found about a 50 percent increase in obesity in C-section babies that could not be explained by other factors. But F. C. Barros et al. ("Cesarean section and risk of obesity in childhood, adolescence, and early adulthood: evidence from 3 Brazilian birth cohorts," *American Journal of Clinical Nutrition* 95 [2012]: 465–70), studying three later birth cohorts, showed effects in the same direction, but they were not statistically significant. The authors discussed unaccounted confounding factors in their study. However, in a later study by the same authors (B. L. Horta et al., "Birth by Caesarean Section and Prevalence of Risk Factors for Non-Communicable Diseases in Young Adults: A Birth Cohort Study," *PLOS ONE* 8 [2013]: e74301), a follow-up of the 1982 birth cohort to the time of their entrance exam into the army at age eighteen (and with follow-up to age twenty-three), they found that C-section was associated with increased body mass index (BMI), amount of body fat, and also systolic blood pressure.

166 **the relationship of C-section with these other modern plagues:** Other disease risks for which some studies have shown increased risk with C-sections: A. K. Hansen et al., "Risk of respiratory morbidity in term infants delivered by elective caesarean section: cohort study," *British Medical Journal* 336 (2008): 85–87; C. Roduit et al., "Asthma at 8 years of age in children born by caesarean section," *Thorax* 64 (2008): 107–13; H. Renz-Polster et al., "Caesarean section delivery and the risk of allergic disorders in childhood," *Clinical and Experimental Allergy* 35 (2005): 1466–72; P. Bager et al., "Caesarean delivery and risk of atopy and allergic disease: meta-analyses," *Clinical and Experimental Allergy* 38 (2008) 634–42; and C. R. Cardwell et al., "Caesarean section is associated with an increased risk of childhood-onset type 1 diabetes mellitus: a meta-analysis of observational studies," *Diabetologia* 51 (2008): 726–35. Not every study shows an association with these diseases; some studies are small and underpowered, and others have multiple confounding factors, but there certainly is mounting evidence that the biological costs of C-section are experienced not just during the first month after birth.

14. MODERN PLAGUES REVISITED

169 **disappearing before their second birthdays:** Type-1 DM epidemiology is changing in the United States: T. H. Lipman et al., "Increasing incidence of type 1 diabetes in youth. Twenty years of the Philadelphia Pediatric Diabetes Registry," *Diabetes Care* 36 (2013): 1597–1603; and in Europe: C. C. Patterson et al., "Incidence trends for childhood type 1 diabetes in Europe during 1989–2003 and predicted new cases 2005–2020: a multicenter prospective registration study," *Lancet* 373 (2009): 2027–33.

170 **babies who gain weight more rapidly in the first year of life:** E. Bonifacio et al., "Cesarean section and interferon-induced helicase gene polymorphisms combine to

increase childhood type I diabetes risk," *Diabetes* 60 (2011): 3300–306; R. M. Viner et al., "Childhood body mass index (BMI), breastfeeding and risk of Type I diabetes: findings from a longitudinal national birth cohort," *Diabetic Medicine* 25 (2008): 1056–61; M. Ljungkrantz et al., "Type I diabetes: increased height and weight gains in early childhood," *Pediatric Diabetes* 9 (2008): 50–56; and E. Hypponen et al., "Obesity, increased linear growth, and risk of type I diabetes in children," *Diabetes Care* 23 (2000): 1755–60. In classic studies of migrants to England, Bodansky and colleagues showed that the children born in the new place (UK) developed higher rates than those born in the old country (H. J. Bodansky et al., "Evidence for an environmental effect in the aetiology of insulin dependent diabetes in a transmigratory population," *British Medical Journal* 304 [1992]: 1020–22). Together, all of these point to strong environmental influences driving the increased rates of Type I diabetes, but farm exposure does not appear to be important (K. Radon et al., "Exposure to farming environments in early life and type I diabetes: a case-control study," *Diabetes* 54 [2005]: 3212–16).

170 **But could anything accelerate it?:** NOD (non-obese diabetic) mice are a particular strain of mice that have enhanced susceptibility to developing autoimmune diabetes, with many characteristics that resemble Type I diabetes in human children. Affected mice have progressive immune-mediated destruction of the Islets of Langerhans, where insulin is produced. The strain was first recognized in Japan in the late 1970s. (See H. Kikutani and S. Makino, "The murine autoimmune diabetes model: NOD and related strains," *Advances in Immunolology* 51 [1992]: 285–322.) Diabetes develops in 50–80 percent of female mice and 20–40 percent of males. Interestingly, when mice are kept in spanking-clean cages and rooms, they get more diabetes. In rooms with a lot of shared bedding, the rates go down. The general observation is that "dirty protects." This suggests that there are transmissible agents (microbes) whose presence affects the risk of diabetes development. The sex difference in the NOD mice differs from that of humans but permits analyses of the relevant factors underlying the dichotomy.

171 **allowed us to include both in the experiments:** Ali and I discussed this. Despite the JDRF limitation, we agreed that we would study both PAT and STAT because we wanted two chances for success. Fortunately I had some other funds from a philanthropist that I could use for the additional expenses, and Ali had received support from the Howard Hughes Medical Institute to advance her career. And with the promise of funding from JDRF, I said to Ali, "Your preliminary work is going so well, instead of just taking one year off from medical school, why don't you take more time off and get a PhD for your work." This would entail a big change in her career plans. I suggested this to her on a Friday. By Monday, she had decided. "I am going for it!" she excitedly told me, and the NYU MD/PhD program immediately accepted her. She has been an incredible student, already making important discoveries.

173 **more than quadrupling since 1950:** Celiac disease is rising in incidence: T. Not et al., "Celiac disease risk in the USA: high prevalence of antiendomysium antibodies in

healthy blood donors," *Scandinavian Journal of Gastroenterology* 33 (1998): 494–98. One in 250 healthy blood donors in the United States: P. H. R. Green et al., "Characteristics of adult celiac disease in the USA: results of a national survey," *American Journal of Gastroenterology* 96 (2001): 126–31. One in 133 adults; 1 in 56, if you include related disorders: J. F. Ludvigsson et al., "Increasing incidence of celiac disease in a North American population," *American Journal of Gastroenterology* 108 (2013): 818–24. One in 141, based on NHANES data: A. Rubio-Tapia, "The prevalence of celiac disease in the United States," *American Journal of Gastroenterology* 107 (2012): 1538–44.

175 **compared to those who didn't:** K. Marild et al., "Antibiotic exposure and the development of coeliac disease: a nationwide case-control study," *BMC Gastroenterology* 13 (2013): 109.

175 **led by Dr. Ben Lebwohl at Columbia University:** B. Lebwohl et al., "Decreased risk of celiac disease in patients with *Helicobacter pylori* colonization," *American Journal of Epidemiology* 178 (2013): 1721–30.

176 **people born by C-section also face an increased risk:** K. Marild et al., "Pregnancy outcome and risk of celiac disease in offspring: a nationwide case-control study," *Gastroenterology* 142 (2012): 39–45.

178 **the risk of developing IBD at an early age:** A. Hviid et al., "Antibiotic use and inflammatory bowel diseases in childhood," *Gut* 60 (2011): 49–54.

178 **in the first year of life:** A. L. Kozyrskyj et al., "Increased risk of childhood asthma from antibiotic use in early life," *Chest* 131 (2007): 1753–59.

179 **as many as one in fifty children has the condition:** S. H. Sicherer et al., "US prevalence of self-reported peanut, tree nut, and sesame allergy: 11-year follow-up," *Journal of Allergy and Clinical Immunology* 125 (2010): 1322–26.

180 **I was able to find records:** L. Hicks et al., "US outpatient antibiotic prescribing, 2010," *New England Journal of Medicine* 368 (2013): 1461–62.

180 **just for those under two:** G. Chai et al., "Trends of outpatient prescription drug utilization in US children, 2002–2010," *Pediatrics* 130 (2012): 23–31.

181 **highest in the states with the highest obesity:** The CDC data on macrolides were first revealed at a meeting (L. Hicks et al., "Antimicrobial prescription data reveal wide geographic variability in antimicrobial use in the United States, 2009," presented at the forty-eighth annual meeting of the Infectious Disease Society of America, Vancouver, Canada, October 21–24, 2010), and the abstract is available online at https://idsa.confex.com/idsa/2010/webprogram/Paper3571.html. In addition to total antibiotic use, the scientists examined both macrolide use and fluoroquinolone use. Fluoroquinolones include ciprofloxacin, levofloxacin, and others. All three maps—total use, macrolides, and fluoroquinolones—look very similar.

I am focusing on macrolides because fluoroquinolones are not often used for children, whereas macrolides are. Azithromycin was the number-two most-prescribed antibiotic in U.S. children in 2010 (see G. Chai et al., "Trends of outpatient prescription drug utilization in US children"). The CDC data do not distinguish between which macrolides are used in which state, but azithromycin is likely to dominate everywhere because of its remarkable growth in sales. A final caveat is that the CDC maps show antibiotic use across all ages. It is not broken down by year of life, so we do not know whether the relationships observed across all ages also hold for children. Such analyses should be done. Obesity levels source: "Overweight and Obesity" (Atlanta: Centers for Disease Control, 2012), accessed http://www.cdc.gov/obesity/data/adult.html.

181 **one in eighty-eight children has autism or autism spectrum disorder:** The generally accepted first written observation of a problem with autism was the paper by Austria-born Leo Kanner, who had established a child psychiatry clinic at Johns Hopkins Hospital (L. Kanner, "Autistic disturbances of affective contact," *Nervous Child* 2 [1943]: 217–50). Since its recognition, there has been much evidence for its rise, despite a trend toward overdiagnosis of autism and its related disorders. See I. Hertz-Picciotto and L. Delwiche, "The rise in autism and the role of age at diagnosis," *Epidemiology* 20 (2009): 84–90; C. J. Newschaffer et al., "The epidemiology of autism spectrum disorders," *Annual Review of Public Health* 28 (2007): 235–58. In 2012, the Centers for Disease Control released an estimate that one in eighty-eight children has an autism-spectrum disorder (http://www.cdc.gov/media/releases /2012/p0329_autism_disorder.html).

182 **can affect cognitive development and mood:** Rodent studies of gut signaling to the brain, involving the microbiome: J. F. Cryan and T. G. Dinan, "Mind-altering microorganisms: the impact of the gut microbiota on brain and behavior," *Nature Reviews Neuroscience* 13 (2012): 701–12; and R. Diaz Heijtz et al., "Normal gut microbiota modulates brain development and behavior," *Proceedings of the National Academy of Sciences* 108 (2011): 3047–52.

182 **the blood of autistic children:** D. Kiser et al., "Review: the reciprocal interaction between serotonin and social behavior," *Neuroscience & Biobehavioral Reviews* 36 (2012): 786–98; and B. O. Yildirim and J. J. L. Derksen, "Systematic review, structural analysis and a new theoretical perspective on the role of serotonin and associated genes in the etiology of psychopathology and sociopathy," *Neuroscience & Biobehavioral Reviews* 37 (2013): 1254–96.

183 **have important bearing on our estrogen status:** We have reviewed this topic (C. S. Plottel and M. J. Blaser, "Microbiome and malignancy," *Cell Host & Microbe* 10 [2011]: 324–35), which includes many of the primary references.

184 **it's not their genes:** M. C. King et al., "Breast and ovarian cancer risks due to inherited mutations in BRCA1 and BRCA2," *Science* 302 (2003): 643–46. According to the U.S. National Cancer Institute, about 12 percent of U.S. women develop breast cancer at some time in their life, but women who have mutations in BRCA1 have a 55 to 65 percent chance, and those with mutations in BRCA2 have a 45 percent chance of developing the cancer before the age of seventy. Ovarian cancer is less

common, but having a BRCA mutation increases the risk even more (lifetime risk in the general population is 1.4 percent; the risk with BRCA1 is 39 percent and the risk with BRCA2 is 11–17 percent). Dr. Mary Claire King was one of the discoverers of BRCA1 and has been a pioneer in studies to understand its significance. In her 2003 review in *Science*, she provided data indicating that women who have these mutations develop breast cancers at different ages. However, what was particularly alarming to me is that among women born after 1940, the age curve had shifted to the left. At any given age, women with BRCA1 or BRCA2 mutations who were born after 1940 had a much higher risk of developing breast cancer than women born before 1940. Although comprehensive genetic analyses were not done, such data suggest that there has been a strong environmental risk added to the genetic risk seen in these women.

184 **as we speculated five years ago:** M. J. Blaser and S. Falkow, "What are the consequences of the disappearing human microbiota?" *Nature Reviews Microbiology* 7 (2009): 887–94.

184 **brilliant *Silent Spring*:** Rachel Carson, *Silent Spring* (New York: Houghton Mifflin, 1962). I read it when I was thirteen. It greatly affected my thinking about the interconnectedness on our planet.

15. ANTIBIOTIC WINTER

185 **a fifty-six-year-old Brooklyn native:** Peggy Lillis's family have started the Peggy Lillis Memorial Foundation to promote public education about *C. diff*.

187 **A recent study of nearly 2 million:** R. E. Polk et al., "Measurement of adult antibacterial drug use in 130 US hospitals: comparison of defined daily dose and days of therapy," *Clinical Infectious Diseases* 44 (2007): 664–70.

188 **lead to greater toxin production:** V. G. Loo et al., "A predominantly clonal multi-institutional outbreak of *Clostridium difficile*–associated diarrhea with high morbidity and mortality," *New England Journal of Medicine* 353 (2005): 2442–49; and M. Warny et al., "Toxin production by an emerging strain of *Clostridium difficile* associated with outbreaks of severe disease in North America and Europe," *Lancet* 366 (2005): 1079–84.

189 **overall picture of drug-resistant bacteria in the United States:** "CDC Threat Report 2013: Antibiotic resistance threats in the United States, 2013," at http://www.cdc.gov/drugresistance/threat-report-2013/.

190 **fending off disease-causing bacteria:** In their initial experiments, Marjorie Bohnhoff and her colleagues showed that the dose of *Salmonella* required to infect half of the exposed mice went from about 100,000 bacterial cells to 3, following a one-day exposure to the antibiotic streptomycin. (M. Bohnhoff et al., "Effect of streptomycin on susceptibility of intestinal tract to experimental *Salmonella* infection," *Proceedings of the Society for Experimental Biology and Medicine* 86 [1954]: 132–37.) In later studies, the team extended the work, showing that penicillin was just as effective as streptomycin, that they could enhance susceptibility of mice to a *Staphylococcus* species that was incapable of colonizing by itself, and that injecting the antibiotic

into tissues had no effect, thus implicating the normal gut bacteria in the protective effect and their depletion by antibiotics in promoting infections. (M. Bohnhoff and C. P. Miller, "Enhanced susceptibility to *Salmonella* infection in streptomycin-treated mice," *Journal of Infectious Diseases* 111 [1962]: 117–27.) These and further observations are more than fifty years old, but they have been largely forgotten.

191 **160,000 people became ill and several died:** C. Ryan et al., "Massive outbreak of antimicrobial-resistant salmonellosis traced to pasteurized milk," *Journal of the American Medical Association* 258 (1987): 3269–74.

192 **found in the human gut and on human skin:** M. Sjölund et al., "Long-term persistence of resistant *Enterococcus* species after antibiotics to eradicate *Helicobacter pylori*," *Annals of Internal Medicine* 139 (2003): 483–87; and M. Sjölund et al., "Persistence of resistant *Staphylococcus epidermidis* after a single course of clarithromycin," *Emerging Infectious Diseases* 11 (2005): 1389–93. *Staphylococcus epidermidis* is a very common type of *Staphylococcus* that colonizes the human skin, and it has much less potential to be a pathogen than *S. aureus*. Change in its abundance is a good indicator of perturbations of the skin environment.

194 **a large number of much less common ones:** Fundamental studies have been done in the last few years describing the outlines of the populations of residential bacteria in our bodies, as well as the genes they carry. For an introduction into this area, see C. Huttenhower et al., "Structure, function and diversity of the healthy human microbiome," *Nature* 486 (2012): 207–14; and J. Qin et al., "A human gut microbial gene catalogue established by metagenomic sequencing," *Nature* 464 (2010): 59–64.

198 **microbial diversity and the genes that accompany it:** T. Yatsunenko et al. found that adults in the United States carried 15–25 percent fewer bacterial species in their intestines than did people who were from Malawi, or were Amerindians in Venezuela, respectively (T. Yatsunenko et al., "Human gut microbiome viewed across age and geography," *Nature* 486 [2012]: 222–27). Le Chattlier and colleagues found that a large proportion of Europeans had about 40 percent fewer bacterial genes than Europeans with a full complement of genes. Those with low gene numbers were much more likely to be obese (E. Le Chatelier et al., "Richness of human gut microbiome correlates with metabolic markers," *Nature* 500 [2013]: 541–46). Although these data are consistent with our idea that depletion of our resident microbes predisposes to obesity (M. J. Blaser and S. Falkow, "What are the consequences of the disappearing human microbiota?" *Nature Reviews Microbiology* 7 [2009]: 887–94), the data do not yet permit ascertaining the direction of causality.

16. SOLUTIONS

199 **I recommended that she start antibiotics immediately:** Lyme disease is caused by *Borrelia burgdorferi*, a bacterium that lives mostly in rodents but can be transmitted by ticks to larger mammals like deer and us.

200 **it kills bacteria on contact:** Triclosan, an antimicrobial and antifungal agent, has been used since the late 1960s to prevent hospital-acquired infections. It was put

into underarm deodorants in the 1970s to reduce microbial populations that contribute to human body smells. Today triclosan is in thousands of products: soaps, toothpaste, pizza cutters, mouthwash, clothing, cleaning supplies, mattresses, and some flooring—anywhere you might want to reduce bacterial or fungal counts. You also see little dispensers of hand sanitizers not only in hospitals but also in grocery stores, offices, classrooms, conference centers, hotels, gyms, in fact, everywhere. As advertisers vilify germs, the public slathers on the triclosan and many products with similar antibacterial effects. Evidence that triclosan is affecting the bacterial communities that live on us is growing. See S. Skovgaard et al., "*Staphylococcus epidermidis* isolated in 1965 are more susceptible to triclosan than current isolates," *PLOS ONE* 16 (2013): e62197; D. J. Stickler and G. L. Jones, "Reduced susceptibility of *Proteus mirabilis* to triclosan," *Antimicrobial Agents and Chemotherapy* 52 (2008): 991–94; and A. E. Aiello et al., "Relationship between triclosan and susceptibilities of bacteria isolated from hands in the community," *Antimicrobial Agents and Chemotherapy* 48 (2004): 2973–79.

201 **are being prescribed annually to U.S. children:** G. Chai et al., "Trends of outpatient prescription drug utilization in U.S. children, 2002–2010," *Pediatrics* 130 (2012): 23–31. Of the top eight drugs given to U.S. children in 2010, five were antibiotics, accounting for more than 41 million individual courses. In steady state, just these five antibiotics would account for about ten courses per child during the first eighteen years of life, and the evidence suggests that we have improved in recent years, so the rate was probably higher in the past. Four of the five are beta-lactam antibiotics, the descendants in a sense of pencillin, and the other was azithromycin, the "Z-pak." Interestingly, the other three drugs in the top-eight list, accounting for 13 million courses, are mostly used for asthma (see chapter 11).

202 **people living in western states:** L. Hicks et al., "US outpatient antibiotic prescribing, 2010," *New England Journal of Medicine* 368 (2013): 1461–62.

203 **the highest rate of antibiotic use:** O. Cars et al., "Variation in antibiotic use in the European Union," *Lancet* 357 (2001): 1851–53. France had more than fourfold higher use than nearby Netherlands.

204 **"only when necessary":** V. Blanc et al., "'Antibiotics only when necessary' campaign in the Alpes-Maritimes District: no negative impact on invasive infections in children in the community 1998–2003," *Presse Med* 37 (2008): 1739–45. Use has fallen by about half (B. Dunais et al., "Antibiotic prescriptions in French day-care centres: 1999–2008," *Archives of Disease in Childhood* 96 [2011]: 1033–37).

204 **And in Sweden:** In response to the U.S. study, Swedish investigators summarized their country's antibiotic use in 2012. The differences are striking. Not only is the aggregate usage less than half (47 percent) of ours, but in the first three years of life, the most crucial period, on average Swedish children are receiving less than one and a half courses of antibiotics versus about four in U.S. children. We are not seeing higher death rates in Swedish children (in fact they are lower), nor more hearing deficits. The regional variation is also less, the difference between the extremes of

urban Stockholm (408/1000) and the rural north (315/1000) about 30 percent. See A. Ternhag and J. Hellman, "More on U.S. outpatient antibiotic prescribing, 2010," *New England Journal of Medicine* 369 (2013): 1175–76. These numbers tell us that major reductions in prescribing can be readily accomplished.

206 **for patients with cancer:** My dad had a low-grade lymphoma diagnosed in his late eighties. He did well on no treatment until about five years later, when he developed a severe form of anemia. Now treatment was needed. After receiving an antibody for a protein on the surface of his malignant cells, he responded immediately and well. In total, he received four weekly injections while sitting in a chair watching TV for a few hours each time. The treatment was fantastic, but the cost, $110,000, was huge. He needed three more of these courses over the next couple of years, and now, nearly five years later, he is fine. He paid his insurance premiums all of those years and made his contributions to Social Security as well. Treatment with this designer drug definitely extended both the quality and quantity of his life. Pharmaceutical companies and hospitals can make large returns this way as long as insurance plans still pay. He has a relatively uncommon condition. But to treat millions of infections in young children with designer drugs of the kind I outlined would break the bank; a different economic is needed.

207 **by identifying specific agents:** X. Hu et al., "Gene expression profiles in febrile children with defined viral and bacterial infection," *Proceedings of the National Academy of Sciences* 110 (2013): 12792–97.

207 **which organism is causing the trouble:** A. Zaas et al., "A host-based RT-PCR gene expression signature to identify acute respiratory viral infection," *Science Translational Medicine* 5 (2013): 203ra126.

207 **even higher than it is in the United States:** L. Dong, "Antibiotic prescribing patterns in village health clinics across 10 provinces of Western China," *Journal of Antimicrobial Chemotherapy* 62 (2008): 410–15. Hospitals can mark up the price of antibiotics sold to patients, providing financial incentives for their overuse. One estimate is that Chinese patients have more than double the antibiotic use of U.S. patients, and on pig farms, it is four times more. In a survey of large pig farms Y.-G. Zhu et al. found 149 different antibiotic-resistance genes, often at extremely high concentrations (Y.-G. Zhu et al., "Diverse and abundant antibiotic resistance genes in Chinese swine farms," *Proceedings of the National Academy of Sciences* 110 [2013]: 3435–40).

209 **no one really knows what causes it:** Diverticulitis is a complication of diverticulosis, a condition with finger-size or smaller out-pouching in the colon. Usually, diverticulosis has no symptoms, and is mostly associated with aging, but occasionally it leads to diverticulitis. As in the patient described, it can be a painful illness with fever, due to inflammation of the wall of the out-pouch.

211 **a much stronger scientific base for their efficacy:** A few probiotics have been successful in treating or preventing infectious diseases. We have limited evidence that probiotics can help prevent *C. diff* infection, and possibly protect against serious infections due to

the particularly virulent *E. coli* (O157:H7) strains (K. Eaton et al., "A cocktail of non-pathogenic bacteria naturally occurring in the digestive tract of healthy humans can protect against a potentially lethal *E. coli* infection [EHEC O157:H7]," abstract presented at the 113th Annual Meeting of the American Society of Microbiology, Denver, CO, May 2013). Eaton and her colleagues gave EHEC to mice colonized by six normal human commensals of the gut or not colonized at all and found that there was no toxin production in the former group but high levels in the latter. These findings suggest possible probiotic candidates for the prevention or treatment of serious EHEC infections.

213 **pivotal and attention-getting study:** Solid evidence that fecal transplantation works to cure patients with recurrent *C. diff* infections (E. van Nood et al., "Duodenal infusion of donor feces for recurrent *Clostridium difficile*," *New England Journal of Medicine* 368 [2013]: 407–15).

213 **not far-fetched to think:** R. A. Koeth et al., "Intestinal microbiota metabolism of L-carnitine, a nutrient in red meat, promotes atherosclerosis," *Nature Medicine* 19 (2013): 576–85; W. H. W. Tang et al., "Intestinal microbial metabolism of phosphatidylcholine and cardiovascular risk," *New England Journal of Medicine* 368 (2013): 1575–84.

214 **ruling was quite reasonable:** A group of physicians and scientists (including myself) was asked by the American Gastroenterological Association to comment on the ruling. Our consensus was that it was appropriate, and we discussed the reasoning and implications. See G. Hecht et al., "What is the value of an FDA IND for fecal microbiota transplantation to treat *Clostridium difficile* infection?" *Clinical Gastroenterology and Hepatology* (2014), in press.

214 **can teach us the key principles?:** I. Pantoja-Feliciano, "Biphasic assembly of the murine intestinal microbiota during early development," *ISME Journal* 7 (2013): 1112–15. Ida, who studied for her PhD with Gloria, examined the relationship of the microbiota in mice in relation to their mother's vagina and intestine. In earliest life, the gut organisms of the pups looked like those of their mother's vagina. While they were nursing, they had a very restricted microbiota dominated by a few major bacteria, like lactobacillus, and then after they were weaned, the profile changed again and resembled that of their mother's intestine. In a few short weeks, Gloria's group had recapitulated the early-life development of the intestinal residents of human children.

215 **Rob Knight and José Clemente:** Rob Knight, a very tall, thin biochemist originally from New Zealand who heads a large research group in Colorado, has been brilliant in the creation of software programs to analyze the complexity of the microbiome and to deconvolute it. José Clemente, originally from Spain, came to work with Rob via Japan and now has his own lab in New York. Rob had come in on a conference and was staying with us. José took the subway down. I was getting ready to take the train up to Brown to speak about my own work. Still I couldn't resist listening in on them, and witnessing the discoveries for myself.

217 **back to our children:** M. J. Blaser, "Science, medicine, and the future: *Helicobacter pylori* and gastric diseases," *British Medical Journal* 316 (1998): 1507–10.

EPILOGUE

219 **ice cap in Greenland would melt:** For a comprehensive view of global warming for
the nonscientist, see, for example: B. E. Johansen, *The Encyclopedia of Global Warming
Science and Technology*, vols. 1 and 2 (Santa Barbara, California: Greenwood Publishing, 2009); and for some solutions: M. Z. Jacobson and M. A. Dilucchi, "A path to
sustainable energy by 2030," *Scientific American* 301 (2009): 58–65.

ACKNOWLEDGMENTS

Writing, as with science, often takes a village, especially when the author, like me, has a different day job. I am much indebted to my daughter Simone Blaser for helping me shape my early ideas into a form that could be attractive to a publisher, and to Dorian Karchmar, my agent (and Simone's boss at William Morris), who helped me get there. Sandra Blakeslee did yeoman work converting my ideas and prose, generated by an academic, into a manuscript that could be more widely understood. To Sandra, with her endless creativity, intellect, and energy, and who I now count as one of the most important teachers in my career, I will be forever grateful. Gillian Blake, editor in chief at Henry Holt, and an enthusiast about this work from the very start, contributed in too many ways to count, and I learned that with regard to both style and content she was always right.

Many of my colleagues read portions of the manuscript to help determine whether or not I was on track and accurate. I appreciate the efforts of Drs. William Ledger, Ernst Kuipers, Claudia Plottel, Gerald Smith, and José Clemente, and the important suggestions of Erika Goldman. Dr. Robert Anderson read the work as both physician and reader, and he gave great advice. I am indebted to Dr. Jan Vilcek for his critical insights as well; although English is not his native language, Jan also corrected my grammar. Linda Peters and Isabel Teitler helped me understand what could be understood and was interesting. I appreciate the friendship they each shared, helping me to craft this manuscript. My assistants at New York University,

Sandra Fiorelli, Jessica Stangel, and then Joyce Ying, helped make order from chaos, no small feat, and I am most appreciative of their efforts. Adriana Pericchi Dominguez was an assiduous and resourceful fact-checker.

An important segment of the book focuses on the research done in my lab at Vanderbilt University and, over the past fourteen years, at NYU. At Vanderbilt, Drs. Tim Cover, Murali Tummuru, Guillermo Pérez-Pérez, Richard Peek, John Atherton, and Ernst Kuipers played key roles. At NYU, it also was very much a team effort, involving other faculty members, graduate and medical students, college and high school students, and visiting researchers. So many were involved in substantive ways that it would difficult to name them all. But for the work highlighted in the text, Drs. Guillermo Pérez-Pérez, Zhiheng Pei, Fritz Francois, Joan Reibman, Yu Chen, Zhan Gao, Ilseung Cho, Claudia Plottel, Alex Alekseyenko, Leo Trasande, and Jan Blustein—all fellow NYU faculty members—contributed in ways mentioned and not. I have had outstanding graduate students and postdoctoral fellows who worked with me on the experiments discussed, notably Laurie Cox, Shingo Yamanishi, Alexandra Livanos, Sabine Kienesberger, and Victoria Ruiz. Yael Noble worked as a research assistant before her time in medical school, but in her efforts she was more like a grad student. Many other students, postdocs, and colleagues are working on ongoing projects that one day will be described in great detail in original scientific publications. Together we have had and continue to have an amazing lab, with a great culture of sharing and generosity.

Hurricane Sandy hit us very hard. With a loss of electrical power, we had a mad dash to retrieve our thawing specimens in freezers— the work of thirty years of research. We rescued nearly all of the current studies, but lost some of our archives—samples obtained from villages and patients all over the world decades ago. They were irreplaceable. We were out of our home lab at the New York Veterans Affairs hospital for more than ten months and had one tribulation piled on the next. Yet with their kindness to one another, adaptability, and "can do" mentality, it was, for the lab members, their finest hour, and the storm and its aftermath provided lessons in life that can not be learned from books.

For the past eight years, my research has had major philanthropic support in the form of the Diane Belfer Program in Human Microbial Ecology. Diane was an early believer in the value of our studies. I much appreciate her enthusiasm and unwavering support, beginning when the ideas were more of a dream. Early support also came from the Ellison Medical Foundation. More recently, the Knapp Family Foundation and the Leslie and Daniel Ziff Foundation have been major sponsors of our explorations. Our work also has been supported by the D'Agostino Foundation, Hemmerdinger Foundation, Fritz and Adelaide Kaufman Foundation, Margaret Q. Landenberger Research Foundation, Graham Family Charitable Foundation, James and Patricia Cayne Trust, and Messrs. David Fox, Richard Sharfman, Michael Saperstein, Robert Spass, and Joseph Curcio, and Dr. Bernard Levine, as well as Mss. Regina Skyer, Edythe Heyman, and Lorraine DiPaolo. Donna Marino has been an incredibly effective advocate for our work. I am very grateful to all.

Our work described in this book has been supported by funding from the National Institutes of Health, the U.S. Army, the Department of Veterans Affairs, the Juvenile Diabetes Research Foundation, the Howard Hughes Medical Institute, the Bill and Melinda Gates Foundation, the Robert Wood Johnson Foundation, the Ellison Medical Foundation, the International Union against Cancer, the World Health Organization, and governments and universities in Japan, the Netherlands, Korea, United Kingdom, Switzerland, Finland, Sweden, France, Italy, Turkey, and Venezuela for support of visiting scholars. Institutional support came in many different forms from the NYU Langone Medical Center and from the Manhattan/ NY Harbor Department of Veterans Affairs Medical Center.

This combination of major research university, U.S. government, private foundations, international support, and philanthropy is necessary for a research program to survive and ultimately to flower.

Finally, my wife and research partner, Dr. Maria Gloria Domínguez Bello, has helped with insight, criticism, adventure, and love. I am glad that I could highlight a few of her many contributions to our shared field. My children Daniel, Genia, and Simone have been steadfast in their love and support.

As with most projects that take a long time, many hands stirred the pot and contributed greatly. I thank one and all for their wonderful help and fellowship.

INDEX